Manufacturing Applications

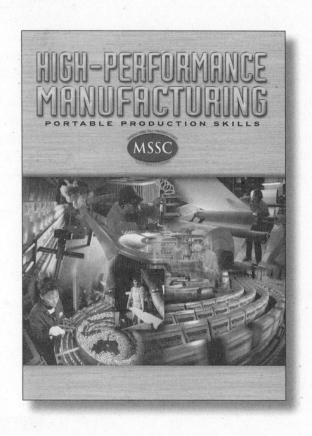

New York, New York   Columbus, Ohio   Chicago, Illinois   Woodland Hills, California

D0985593

Safety Notice

The reader is expressly advised to consider and use all safety precautions described in this book or that might also be indicated by undertaking the activities described herein. In addition, common sense should be exercised to help avoid all potential hazards.

Publisher assumes no responsibility for the activities of the reader or for the subject matter experts who prepared *Manufacturing Applications*. Publisher makes no representation or warranties of any kind, including but not limited to, the warranties of fitness for particular purpose or merchantability, nor for any implied warranties related thereto, or otherwise. Publisher will not be liable for damages of any type, including any consequential, special or exemplary damages resulting, in whole or in part, from reader's use or reliance upon the information, instructions, warnings or other matter contained in *Manufacturing Applications*.

Brand Disclaimer

Publisher does not necessarily recommend or endorse any particular company or brand name product that may be discussed or pictured in this text. Brand name products are used because they are readily available, likely to be known to the reader, and their use may aid in the understanding of the text. Publisher recognizes that other brand name or generic products may be substituted and work as well or better than those featured in *Manufacturing Applications*.

Acknowledgment

Publisher gratefully acknowledges Albuquerque TVI Community College for use of material on pages 67–68, 117–118, and 265.

The McGraw-Hill Companies

Copyright © 2006 by The McGraw-Hill Companies, Inc.
All rights reserved. Except as permitted under the United States Copyright Act, no part of this publication may be reproduced or distributed in any form or by any means, or stored in a database or retrieval system, without prior written permission of the publisher, Glencoe/McGraw-Hill.

Send all inquiries to:
Glencoe/McGraw-Hill
21600 Oxnard Street, Suite 500
Woodland Hills, CA 91367

ISBN 0-07-861183-0
Printed in the United States of America
2 3 4 5 6 7 8 9 10   009   11 10 09 08 07

Contents

There are no application activities for Chapters 1–3 of the textbook.

Chapter 4 Safety Practices

Chapter 5 Communication Skills

Chapter 6 Teamwork Skills

COPYRIGHT © GLENCOE/McGRAW-HILL

Chapter 11 Tool & Equipment Operation

Chapter 12 Production Planning & Work Flow

Chapter 13 Production Components

Chapter 14 Controlling & Documenting Production

Chapter 15   Packaging & Distributing Products

Chapter 16   Continuous Improvement

Chapter 17   Inspection & Auditing

Chapter 18   Preventive & Corrective Actions

   COPYRIGHT © GLENCOE/McGRAW-HILL

Posting Worker Rights

The Department of Labor (DOL) and some state government agencies require employers to post notices in the workplace to inform workers of their rights. Employers are required to inform employees about their safety and health protections under the Occupational Safety and Health Act, which is administered by the Occupational Safety and Health Administration (OSHA). Employers must also post the Equal Employment Opportunity Act, which protects against employment discrimination.

What Must Be Posted

Worker rights posters are required by law for employers who meet certain requirements. They must be posted in a conspicuous place that is accessible to all employees and job applicants. For your own protection, look for these posters at your workplace and be sure that you understand the information. If you don't know where to find them, ask your supervisor. In addition to informing you about the protections of the particular law, the poster provides contact information for assistance.

The following posters should be found in your workplace.

- Job Safety and Health Protection—the official OSHA poster, approved by the DOL. This poster states that you have a right to a safe and healthful workplace.

- Equal Employment Opportunity. Employers may not discriminate against employees or job applicants.

- Federal Minimum Wage—information about minimum wage and overtime pay requirements.

- Family and Medical Leave—information about the right to unpaid leave without penalty for certain medical and family needs.

- State-required posters concerning state labor laws.

- Other posters that may be required, when applicable, concerning youth employment, migrant workers, and minimum wage for disabled workers.

Protecting Workers

Workers today have many more rights than they did long ago. Strong efforts by labor unions, demands by civil rights leaders, and work by public interest

groups have led to major changes in U.S. labor laws. These laws include the Occupational Safety and Health Act and the Equal Employment Opportunity Act. They guarantee employees the right to safe, healthy, and fair working conditions. Other laws include:

Civil Rights Act. It prohibits discrimination based on race, color, national origin, sex, or religion. It protects U.S. citizens working for U.S. companies in other countries.

(continued on next page)

Equal Pay Act. It requires that men and women be given equal pay for substantially equal work in the same establishment.

Age Discrimination in Employment Act. It protects people 40 years of age and older from being discriminated against in any aspect of employment.

Americans with Disabilities Act. It prevents employers from refusing to hire or promote disabled persons and ensures that all employees are treated equally.

Immigration Reform and Control Act. Only U.S. citizens and people who are authorized to work in the United States may be legally hired.

Federal Employment Compensation Act. This act directs states to regulate workers' compensation programs to insure employees against injuries that occur on the job.

APPLYING YOUR KNOWLEDGE

1. Which federal government department requires that employers place posters concerning regulations in the workplace?

2. What kinds of posters are required at the workplace?

3. Where should the posters be displayed?

PRODUCTION CHALLENGE

You have been appointed the information officer for a manufacturer that has just opened a new plant. In this role, you need to ensure that the appropriate state and federal regulations are posted. These regulations are not currently posted. What can you do to ensure that you are posting the correct regulations in the correct places within the plant?

MEETS THE FOLLOWING PRODUCTION STANDARDS

P1-07B	Knowledge of state and federal regulatory requirements (e.g., OSHA).
P7-01G	Knowledge of what the law requires companies to post or publish in order to keep employees abreast of OSHA and other government regulations.

COPYRIGHT © GLENCOE/McGRAW-HILL

Conducting Safety Orientation & Training

The risk of injury is much greater for new workers than it is for those with more experience. According to one government study, workers on the job for less than one year were 40% more likely to be injured than other workers. In most cases, the newer workers lacked basic accident prevention and safety knowledge. To reduce these risks, employers use various safety orientation and training techniques.

You might be the person who provides this training. You should demonstrate openness to new safety procedures. When interacting with new employees, stress the importance of a safe work environment. Stress the importance of teamwork. Building teamwork skills can help build a common theme of "safety first" among workers. This can help to ensure a safe work environment.

A plant tour can be a key part of the orientation. It is important in helping workers understand dangers and prepare for potential emergencies. You should point out emergency exits, evacuation routes, and hazardous areas.

New worker training is essential for safe manufacturing. You will need to develop skill in conducting equipment safety demonstrations. Each frontline worker should learn how to use machines and tools safely. Training on the use of new machines will expand their skill base. Maintenance workers should obtain certification to train others in technical skills and knowledge, where applicable. This will help minimize safety issues.

Safety Orientation

When you conduct a safety orientation, communicate clearly to ensure that safety issues are understood and safety practices used. Make sure that workers stay focused and ask questions as needed. The safety orientation should cover:

❑ Company first-aid and first-response procedures.

❑ Location and use of emergency alarms.

❑ Emergency procedures, including evacuation routes and emergency exits.

❑ Location and use of first-aid kits.

❑ Location and operation of safety equipment, such as fire extinguishers.

❑ How to inspect a work area and report possible safety risks.

❑ Possible hazards in the workplace to help ensure personal safety as well as the safety of others.

❑ How to handle hazardous materials if this is part of the new worker's job.

❑ How to report hazards.

❑ Health and safety standards to ensure that quality problems are addressed correctly without impairing health and safety.

❑ The need for employees to raise safety concerns, ask questions, and receive additional training.

❑ The processes for employees to raise safety concerns, ask questions, and receive additional training.

❑ How to report accidents and incidents.

❑ Proper PPE to be worn.

❑ Identification and use of color codes, alarms, signs, labels, and other devices that warn of hazards.

❑ The location of MSDS if the worker's job involves exposure to hazardous materials.

❑ The location of posters that inform workers of the federal and state regulations concerning worker rights and protections.

❑ The location of the OSHA posters presenting safety information.

(continued on next page)

Additional Safety Training

All workers will from time to time need additional safety training. Frequent refresher courses are essential in reinforcing workplace safety practices. You should anticipate the need for training on the safe use of equipment. As new machines or processes are added, more training may be required. Information must be gathered to identify which workers need safety training. You can help determine the frequency of safety training and build consensus in the workplace on what level of safety training is needed. You can also plan safety-related training based on information received from equipment installers, maintenance technicians, and operators. Monthly meetings can be used to improve the safety environment and communicate changes in regulations. Safety training must be delivered regularly.

Documenting Safety Orientation & Training

Following training, workers must remain informed and up-to-date. The orientation and training of employees must be documented according to the company's requirements. Most companies require that training records be kept. The records must describe the orientation and training and provide the names of participants and trainers and the dates of training. Most likely, you will use a computer to track safety training. The records would then be stored electronically.

MEETS THE FOLLOWING (MSSC) PRODUCTION STANDARDS

P2-A11C	Attend training on new machines to improve skill base.
P3-K4	Provide safety orientation training for other employees.
P3-K4A	Orientation covers all topics and procedures needed to facilitate employee safety.
P3-K4B	Orientation makes clear the need and processes for employees to raise safety concerns, ask questions, and receive additional training.
P3-K4C	Orientation is documented according to company requirements.
P3-K4D	Safety training is delivered regularly.
P3-O5A	Skill in developing and/or delivering safety training per guidelines.
P3-A1E	Use computer to track safety training.
P3-A2D	Gather information on who is in need of safety training.
P3-A4D	Determine the frequency of safety training and drills.
P3-A6C	Interact with new employees on importance of safe work environment in order to make a positive impact.
P3-A9E	Build a common theme of "safety-first" among workers to ensure a safe work environment.
P3-A10E	Build consensus on what level of safety training is needed.
P3-A14E	Document safety incident and training orientation.
P5-K1A	Communication is sufficient to ensure that safety issues are understood and safety practices used.
P7-K1C	Maintenance workers obtain certification to train others in technical skills and knowledge, where applicable.
P7-O3D	Skill in conducting equipment safety demonstrations.
P7-O3G	Skill in using monthly safety meetings to improve the safety environment and communicate changes in regulations.
P7-A3A	Anticipate, identify, and provide responsive preventative training for safe use of equipment.
P7-A5A	Plan safety-related training for equipment based on operator maintenance, installer experience.
P7-A7C	Demonstrate openness to new safety procedures.

 COPYRIGHT © GLENCOE/MCGRAW-HILL

Conducting Walk-Through Inspections

Workplace inspections are needed to ensure safe working conditions. One type of workplace inspection is the walk-through inspection. In a walk-through inspection, the inspector inspects a certain area of the workplace using established safety audit processes. The inspector makes a judgment about improving processes to reduce or eliminate safety injuries.

Purpose

The purpose of a walk-through inspection is:

- To identify unsafe acts and conditions.
- To identify the need for specific equipment guards.
- To communicate safety habits to supervisors and employees.
- To improve safety practices.
- To assist supervisors in inspections.
- To monitor equipment and operator performance, especially as it relates to the careful observance of safety practices.
- To prompt a closer working relationship between safety personnel, supervisors, and employees.

Preparation

Prepare for a walk-through inspection as follows:

- ❑ Clearly define the inspection area and plan the inspection route.
- ❑ Review notes from previous inspections and the log of occupational injuries and illnesses.
- ❑ Ensure that you are knowledgeable regarding the appropriate safety standards.
- ❑ Read pertinent available information regarding the equipment to be inspected.
- ❑ Read pertinent available information regarding proper and safe equipment installation techniques as described in manuals, checklists, and regulations.
- ❑ Make up the appropriate checklists.

General Inspection

The value of a walk-through inspection depends on the inspector's skill in recognizing and reporting unsafe conditions. Follow these guidelines:

- ❑ Limit the inspection to the time allotted.
- ❑ Conduct the inspection while employees are working. This will help ensure that unsafe conditions will be detected and corrected.
- ❑ Do not be a disturbing influence. Try to minimize work interruptions.
- ❑ Document the inspection and record all pertinent data. Take detailed notes. Include the hazard location, name of employee, witnesses, and any comments.
- ❑ Analyze processes to ensure that safety standards have been met.
- ❑ Be alert for all hazards.
- ❑ Check for all potentially unsafe conditions and practices.
- ❑ Check that workers are observing safe tolerances in the workplace.
- ❑ Check for out-of-compliance conditions.
- ❑ Determine why an unsafe condition exists.
- ❑ Ask employees to identify any concerns regarding safety or working conditions.
- ❑ Ask operators about existing or potential problems regarding the equipment.
- ❑ Identify the need for on-the-spot corrections of unsafe practices or procedures.

(continued on next page)

Using Safety Inspection Checklists

A safety inspection checklist must be tailored to the industry and the individual workplace. Through experience, you can develop skill in preparing safety checklists. The following are examples of what you might find on a complete safety inspection checklist.

First-Aid

- ☐ First-aid supplies and equipment available at all times.
- ☐ Eyewash stations available and working properly.
- ☐ Emergency showers available and working properly.
- ☐ Certified first responders available.

Personal Protective Equipment

- ☐ Required personal protective equipment (PPE) provided and required.
- ☐ Training on proper use of PPE.
- ☐ Proper PPE worn.
- ☐ Respirators available and worn.

Sanitation

- ☐ Proper housekeeping.
- ☐ Cleanup supplies available (soap, towels).
- ☐ Hot or tepid water available.

Storage

- ☐ Storage properly stacked and secured.
- ☐ Chemicals stored properly.
- ☐ Clearance capacity posted for overhead storage.
- ☐ Aisles marked and kept clear.

Walking Surfaces

- ☐ All clean and dry.
- ☐ Adequate illumination.
- ☐ Slip-resistant mats in place.
- ☐ Wet floor signs in use.
- ☐ Aisles marked and kept clear.
- ☐ Stairways sturdy with adequate railing.
- ☐ Stairways kept clear and dry.

Exits

- ☐ Marked with easily seen signs.
- ☐ Directional exit signs.
- ☐ Free of obstructions.
- ☐ All exit doors labeled.

Fire Protection

- ☐ Fire hazards removed or guarded.
- ☐ Fire extinguishers available and checked.
- ☐ Emergency alarms operating properly.

Ventilation

- ☐ Mechanical ventilation provided as necessary.
- ☐ Ventilation adequate to remove air contaminants.
- ☐ Exhaust systems provide adequate ventilation.

Machine Guards

- ☐ Safety guards are not bypassed.
- ☐ Grinding wheels guarded.
- ☐ Flywheels guarded.
- ☐ Pulleys guarded.
- ☐ Belt drives guarded.
- ☐ Sprocket wheels and chains guarded.
- ☐ Saw blades guarded.
- ☐ Jointers guarded.
- ☐ Rotating shafts guarded.
- ☐ Fan blades guarded.

Portable Power Tools

- ☐ Portable power tools properly guarded.
- ☐ Portable power tools properly grounded.

Electrical

- ☐ Electrical wiring complies with National Electrical Code.
- ☐ Electrical equipment properly grounded.
- ☐ Electrical cords protected from damage.
- ☐ Extension cords not used as fixed wiring.
- ☐ Breakers properly labeled.

(continued on next page)

COPYRIGHT © GLENCOE/MCGRAW-HILL

Hazard Communication

❏ Containers are properly labeled.

❏ MSDS for all chemicals.

❏ Proper HAZCOM training.

Hazardous Materials

❏ Compressed-gas cylinders labeled and stored properly.

❏ Flammables and combustibles stored properly.

❏ "No Smoking" signs in hazardous areas.

Workstations & Equipment

❏ Workstations clear of safety hazards.

❏ Manufacturer's instructions and rules followed.

❏ Equipment periodically inspected and records maintained.

❏ Trained and certified operators only.

Lockout/Tagout

❏ Proper procedures followed.

❏ Employees notified.

❏ Training provided.

❏ Hardware and tags available and used.

❏ Regular inspections.

Follow-Up

❏ Discuss corrective actions with supervisors and employees in a timely way.

❏ Report issues and problems effectively. Ensure that your recommendations for corrective action are clear, concise, and supported by data.

❏ Ensure that all data was reviewed prior to making recommendations.

❏ Recommend corrective actions for out-of-compliance conditions and unsafe conditions according to proper procedures and documentation.

❏ Recommend corrective actions for quality issues impacting the health or safety of workers according to proper procedures and documentation.

❏ Follow up on corrective actions recommended as a result of your inspection.

❏ File and store inspection documents according to standard company policy.

(continued on next page)

MEETS THE FOLLOWING (MSSC) PRODUCTION STANDARDS

P2-K4C	Unsafe conditions are identified and reported promptly.
P2-K4D	Corrective action is taken to correct unsafe conditions.
P2-O4F	Skill in identifying and reporting unsafe work conditions.
P2-A2B	Perform a walk-through equipment inspection to monitor equipment.
P3-K1	Perform environmental and safety inspections.
P3-K1A	Potential hazards in the work are identified, reported, monitored.
P3-K1D	Inspections meet all relevant, health, safety, and environmental laws and regulations.
P3-K1E	Inspections are done according to company schedule and procedures.
P3-K1F	Inspections are documented.
P3-K1G	Inspection records are stored correctly.
P3-K3A	Conditions that present a threat to health, safety and the environment are identified, reported, and documented promptly.
P3-O2A	Skill in identifying and reporting unsafe conditions.
P3-O2D	Skill in determining if all safety guards are in place prior to machine operation.
P3-O3A	Knowledge of basic filing procedures to properly store inspection records.
P3-O3B	Knowledge of safety requirements and environmental regulations related to performing inspections.
P3-A2B	Visually inspect work area for possible safety hazards.
P3-A4B	Determine that all safety equipment and guards are in place.
P3-A5B	Plan and organize safety and environmental inspections in order to prevent accidents.
P3-A6B	Interact with peers to share information on emergency drills/procedures.
P3-A16C	Measure the distances needed to maintain safe tolerances in the workplace.
P4-K3C	Suggestions are made according to proper procedures and documentation.
P4-K3D	Suggestions show that all data was reviewed prior to making recommendation.
P7-K2	Suggest process and procedures that support safety and effectiveness of work environment.
P7-K2D	Suggestions are properly documented.
P7-K3E	Equipment is audited to ensure there are no by-passes of safety guards.
P7-K4	Monitor equipment and operator performance.
P7-K4B	Out-of-compliance or unsafe conditions are reported immediately.
P7-K4G	Information in the equipment use is gathered from operators to reveal existing or potential problems.
P7-O2D	Skill in regularly monitoring equipment for unsafe conditions.
P7-O4F	Skill in evaluating workplace safety using safety audit processes.
P7-O4G	Knowledge of hazard to document and communicate corrective actions and monitor performance.
P7-O4I	Knowledge of proper and safe installation techniques as described in manuals, checklists and regulations.
P7-O4J	Skill in recognizing and proposing ways to improve safety practices to propose alternative practices.
P7-A4B	Make a judgment about improving processes to reduce or eliminate safety injuries.
P7-A9B	Perform on the spot corrections of unsafe practices and procedures.
P7-A11F	Improve safe working conditions.
P8-K2C	Recommendations for action are clear, concise, and supported by data.
P8-K2D	Recommendations are made to the appropriate parties in a timely way.
P8-K2E	Follow-up activities indicate that corrective action was taken.
P8-K5B	Corrective action is taken for quality issues impacting the health or safety of workers.
P8-O1I	Skill in recognizing and reporting unsafe conditions.

    COPYRIGHT © GLENCOE/McGRAW-HILL

Conducting Environmental Testing

The environment includes the air we breathe, the water we drink and use, and the land and structures in which we live and work. Environmental testing detects the presence of hazardous substances in the environment. The effects of exposure to any hazardous substance depend on the dose, the duration, how you are exposed, your personal traits and habits, and whether other chemicals were present. Indoor environmental quality refers to the quality of the environment inside a building. Indoor environmental quality (IEQ) problems are preventable and solvable. Practical guidance on how to manage your building for good indoor environmental quality is available.

Properly conducted environmental inspections are essential to the maintenance of a safe workplace environment. Conduct a thorough review of health, safety, and environmental documentation and policies on a regular basis. Ensure that inspections meet all relevant health, safety, and environmental laws and regulations. Perform inspections according to company schedule and procedures. Plan and organize environmental inspections in order to prevent accidents.

The purpose of environmental testing is to identify potential hazards in the workplace. The checklists that follow identify some items for which workplaces are commonly tested. Testing is often done by a government agency such as the National Institute for Occupational Safety and Health (NIOSH).

Air Quality

"Sick Building Syndrome" is a term sometimes used to convey a wide range of symptoms that some think can be caused by the building itself. Poor air quality inside the building is thought to contribute to this syndrome. The workplace environment may be the cause if workers' symptoms are alleviated when they leave the building.

❑ Check for indoor environmental contaminants.
❑ Check for environmental contaminants drawn in from outdoors.
❑ Check for ventilation system deficiencies.
❑ Check for problems with air quality resulting from overcrowding.
❑ Check for problems with air quality resulting from offgassing from materials in the building.

❑ Check for problems with air quality resulting from mechanical equipment.
❑ Check for problems with air quality resulting from tobacco smoke.
❑ Check for problems with air quality resulting from microbiological contamination.
❑ Review environmental data systems in the areas to be inspected.

Carbon Monoxide

Carbon monoxide, or CO, is an odorless, colorless gas that can cause sudden illness and death. It is found in combustion fumes, such as those produced by cars and trucks, small gasoline engines, and heating systems.

❑ Check that carbon monoxide alarms have been installed.
❑ Check that fuel-burning appliances are properly installed, maintained, and operated.
❑ Check that furnaces and water heaters are inspected annually by a qualified service technician.

Hazardous Substances

Hazardous substances are those substances that pose a significant potential threat to human health due to their known or suspected toxicity and potential for human exposure. These substances include arsenic, lead, mercury, vinyl chloride, and white phosphorus.

❑ Check for the presence of hazardous substances in the workplace.
❑ Check that hazardous substances are properly identified.

(continued on next page)

❑ Check that hazardous substances are properly stored and handled.

❑ Check that hazardous substances are properly disposed of.

Hydrocarbons

In a manufacturing environment, exposure to polycyclic aromatic hydrocarbons (PAHs) comes from breathing air containing PAHs in smoke-houses and in coking, coal-tar, and asphalt production plants. Some PAHs can cause cancer.

❑ Check air quality to detect the presence of PAHs.

Lead

Lead does not break down or decompose. Therefore, lead from products such as old paints and discarded batteries remains in the environment. Exposure to lead should be avoided.

❑ Check for the presence of lead in airborne particles, dust, paint, and soil.

Molds

Molds can cause severe reactions among workers exposed to large amounts in occupational settings. Occupational sites with high mold exposures include greenhouses and mills.

❑ Check for the presence of molds.

Follow-Up Procedures

At the conclusion of the inspection, certain actions must be taken if the inspection is to be effective. The appropriate parties should follow these guidelines at the conclusion of the inspection.

❑ Report potential hazards in the workplace.

❑ Take corrective action to correct potential hazards in the workplace.

❑ Monitor potential hazards in the workplace.

❑ Correctly document safety and environmental inspections.

❑ Correctly store the inspection records.

MEETS THE FOLLOWING (MSSC) PRODUCTION STANDARDS

P3-K1	Perform environmental and safety inspections.
P3-K1A	Potential hazards in the work are identified, reported, monitored.
P3-K1B	Corrective action is taken to correct potential hazards.
P3-K1C	Health, safety and environmental documentation and policies are thoroughly and regularly reviewed.
P3-K1D	Inspections meet all relevant, health, safety, and environmental laws and regulations.
P3-K1E	Inspections are done according to company schedule and procedures.
P3-K1F	Inspections are documented.
P3-K1G	Inspection records are stored correctly.
P3-A5B	Plan and organize safety and environmental inspections in order to prevent accidents.
P7-K3D	Environmental testing of workplace is performed on a regular basis as required by company policy or regulation.
P7-O2B	Skill in reviewing environmental data systems in the factory.

COPYRIGHT © GLENCOE/McGRAW-HILL

Writing Accident, Incident & Near Miss Reports

Even when safety practices are followed in the workplace, accidents can happen. They can include worker injury, equipment failure, equipment damage, a near miss—a serious accident that was narrowly avoided—or even a fatality. When these events occur, the first-line supervisor must immediately be notified. Anyone who is injured must be treated or sent to a medical facility.

It's important to report an accident, incident, or near miss to:

- Determine the cause.
- Prevent it from happening again.
- Improve health and safety in the workplace.
- Comply with company and regulatory procedures.
- Gather, analyze, and compare present safety conditions to past safety conditions.
- Forward data to appropriate personnel for inclusion in OSHA recordables.

If you have been involved in an accident, incident, or near miss, or if you have witnessed one, you may need to fill out a report form. In some cases, the first-line supervisor or a safety officer will fill out the form. These forms vary from organization to organization, but they will always require the following information:

- Who was involved.
- What exactly happened.
- Where exactly the event occurred.
- When exactly the event occurred.
- Why it occurred, including direct and indirect causes.
- How it occurred, i.e., the sequence of events.

They might also ask if proper PPE was worn, if SOPs were followed, if workers were trained properly, or if such an incident would be likely to occur again. In addition, you might need to describe steps that could be taken to prevent such an occurrence from happening again. These could include:

- Additional worker training.
- Equipment modification.

- Additional PPE.
- Monitoring equipment.
- Changes in SOPs.
- Corrective action procedures.

Sharing the Report

This report must be tracked and documented as appropriate. Accident, incident, and near miss reports must be completed and forwarded to the appropriate parties promptly—within 24 hours of the occurrence. It is likely that your employer will have such a form on their network or Intranet site. In that case, the form can be completed on the computer and forwarded electronically to the appropriate parties.

In addition to individuals, departments, and possibly regional or national offices within the company, these reports are shared with OSHA and other government offices. For instance, if an incident involves a chemical spill, the EPA must be notified. Sometimes state and local agencies and offices must be notified too.

Completing the Form

The form that follows is an accident investigation report. Note that it can also be used to report a near miss. Complete the form as if an accident or near miss has occurred. Be specific, descriptive, and honest when answering the questions. The report will be used to tabulate safety incidents. Gathering and tracking such information provides valuable information. This can be used to determine how the workforce can be educated to use equipment more safely. Remember that these reports can help prevent such occurrences in the future by establishing what exactly happened.

(continued on next page)

Accident Investigation Report

Instructions: Complete this form as soon as possible after an accident that results in serious injury or illness.
(Optional: Use to investigate a minor injury or near miss that *could have resulted in a serious injury or illness*.)

This is a report of a:	☐ Death ☐ Lost Time ☐ Dr. Visit Only ☐ First Aid Only ☐ Near Miss
Date of incident:	This report is made by: ☐ Employee ☐ Supervisor ☐ Team ☐ Final Report

Step 1: Injured employee (complete this part for each injured employee)

Name:	Sex: ☐ Male ☐ Female	Age:
Department:	Job title at time of accident:	

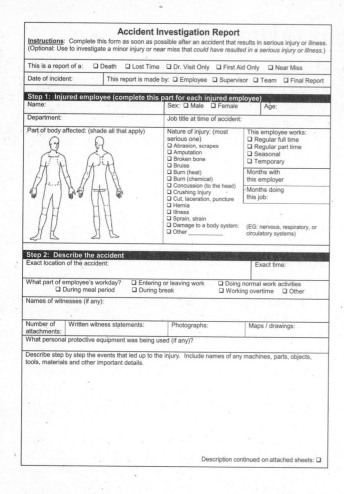

Part of body affected: (shade all that apply)

Nature of injury: (most serious one)
- ☐ Abrasion, scrapes
- ☐ Amputation
- ☐ Broken bone
- ☐ Bruise
- ☐ Burn (heat)
- ☐ Burn (chemical)
- ☐ Concussion (to the head)
- ☐ Crushing Injury
- ☐ Cut, laceration, puncture
- ☐ Hernia
- ☐ Illness
- ☐ Sprain, strain
- ☐ Damage to a body system:
- ☐ Other _____
(EG: nervous, respiratory, or circulatory systems)

This employee works:
- ☐ Regular full time
- ☐ Regular part time
- ☐ Seasonal
- ☐ Temporary

Months with this employer

Months doing this job:

Step 2: Describe the accident

Exact location of the accident:

Exact time:

What part of employee's workday?
- ☐ Entering or leaving work
- ☐ During meal period
- ☐ During break
- ☐ Doing normal work activities
- ☐ Working overtime ☐ Other

Names of witnesses (if any):

Number of attachments:	Written witness statements:	Photographs:	Maps / drawings:

What personal protective equipment was being used (if any)?

Describe step by step the events that led up to the injury. Include names of any machines, parts, objects, tools, materials and other important details.

Description continued on attached sheets: ☐

Accident Investigation Report

Step 3: Why did the accident happen?

Unsafe workplace conditions: (Check all that apply)
- ☐ Inadequate guard
- ☐ Unguarded hazard
- ☐ Safety device is defective
- ☐ Tool or equipment defective
- ☐ Workstation layout is hazardous
- ☐ Unsafe lighting
- ☐ Unsafe ventilation
- ☐ Lack of needed personal protective equipment
- ☐ Lack of appropriate equipment / tools
- ☐ Unsafe clothing
- ☐ No training or insufficient training
- ☐ Other: _____

Unsafe acts: (Check all that apply)
- ☐ Operating without permission
- ☐ Operating at unsafe speed
- ☐ Servicing equipment that has power to it
- ☐ Making a safety device inoperative
- ☐ Using defective equipment
- ☐ Using equipment in an unapproved way
- ☐ Unsafe lifting by hand
- ☐ Taking an unsafe position or posture
- ☐ Distraction, teasing, horseplay
- ☐ Failure to wear personal protective equipment
- ☐ Failure to use the available equipment / tools
- ☐ Other: _____

Why did the unsafe conditions exist?

Why did the unsafe acts occur?

Is there a reward (such as "the job can be done more quickly", or "the product is less likely to be damaged") that may have encouraged the unsafe conditions or acts? ☐ Yes ☐ No
If yes, describe:

Were the unsafe acts or conditions reported prior to the accident? ☐ Yes ☐ No

Have there been similar accidents or near misses prior to the accident? ☐ Yes ☐ No

Step 4: How can future accidents be prevented?

What changes do you suggest to prevent this accident/near miss from happening again?
- ☐ Stop this activity
- ☐ Guard the hazard
- ☐ Train the employee(s)
- ☐ Train the supervisor(s)
- ☐ Redesign task steps
- ☐ Redesign work station
- ☐ Write a new policy/rule
- ☐ Enforce existing policy
- ☐ Routinely inspect for the hazard
- ☐ Personal Protective Equipment
- ☐ Other: _____

What should be (or has been) done to carry out the suggestion(s) checked above?

Description continued on attached sheets: ☐

Step 5: Who completed and reviewed this form? (Please Print)

Written by:	Title:
Department:	Date:

Names of investigation team members:

Reviewed by:	Title:
	Date:

MEETS THE FOLLOWING (MSSC) PRODUCTION STANDARDS

P3-K2C	Emergency drills and incidents are documented promptly according to company and regulatory procedures.
P3-O1D	Knowledge of how to be proactive in responding to a safety concern and document occurrences.
P3-O4C	Knowledge of accident documentation procedures.
P3-A2A	Gather, analyze, and compare present safety conditions to past.
P3-A14B	Write accurate accident injury reports.
P3-A16B	Tabulate safety incidents.
P5-K1H	Communications are tracked and documented as appropriate.
P5-O3C	Knowledge of company reporting forms and documents and procedures specific to safety.
P7-K4F	Accident and injury data is forwarded to appropriate personnel for inclusion in OSHA recordables.
P7-O4N	Skill in generating and sharing near miss reports.
P7-A2E	Gather and track safety metrics so that the workforce can be educated on how better they may utilize equipment safety.
P7-A14A	Create detailed near miss reports to educate workers.

COPYRIGHT © GLENCOE/McGRAW-HILL

Selecting Personal Protective Equipment

Many manufacturers use the "Hierarchy of Health and Safety Controls" to ensure a safe workplace. One level of the hierarchy requires employers to provide personal protective equipment (PPE) to protect workers against different types of hazards. Employers must provide PPE to workers working in areas that might cause injuries or health-related problems. PPE includes safety glasses, ear plugs, face shields, and safety harnesses.

The following checklist can be used to ensure that proper PPE is provided and used.

☐ Employers assess the workplace to determine if hazards that require the use of PPE (e.g., head, eye, face, hand, or foot protection) are present or are likely to be present.

☐ Correct, appropriate, and properly fitted PPE is supplied by the employer where hazards cannot be controlled in any other way.

☐ All PPE is of safe design and construction for the work to be performed.

☐ PPE is checked and maintained regularly so it is sanitary and ready for use.

☐ PPE is used in the correct manner.

☐ PPE is worn by workers when it is required.

Employers must also provide training in the proper use and wear of PPE.

☐ Training is provided to each worker who is required to use PPE.

☐ Each affected worker demonstrates an understanding of the training and the ability to use PPE properly before being allowed to perform the work requiring the use of PPE.

Workers are trained to know at least the following:

☐ When PPE is necessary.

☐ What PPE is necessary.

☐ How to properly put on, remove, adjust, and wear PPE.

☐ The limitations of the PPE.

☐ The proper care, maintenance, useful life, and disposal of the PPE.

Eye Protection

Approximately 1,000 eye injuries occur every day in the workplace. More than half of workers injured in this way are not wearing eye protection. Flying or falling objects or sparks striking the eye cause 70% of eye injuries. Another 20% are caused by contact with chemicals. Face shields with glasses or goggles help protect workers from hazards such as welding sparks, particles and grinding debris, and splashed chemicals.

☐ Protective goggles or face shields are provided and worn where there is any danger of flying particles, corrosive or explosive materials, molten metals, or sparks of any kind. They should also be worn for heat treating operations or furnace work.

☐ Approved safety glasses are required to be worn at all times in areas where there is a risk of eye injuries such as punctures, abrasions, contusions, or burns.

☐ Workers who need corrective lenses (glasses or contacts) in working environments with harmful exposures are required to wear only approved safety glasses, protective goggles, or use other medically approved precautionary procedures.

☐ Safety glasses meet both industry and OSHA specifications.

☐ Eye protective equipment is provided to each worker for hygienic reasons. If it is necessary to share this equipment, it should be thoroughly cleaned and disinfected before another worker uses it.

(continued on next page)

Ear Protection

Some tools and machines, such as high-speed drills, can be very noisy. At certain levels over a period of time, such noise can cause permanent hearing loss. Meters are often used to monitor noise levels in work areas. When levels are elevated, earplugs, earmuffs, or both can be worn for hearing protection.

❑ Employers provide ear plugs or ear muffs against the effects of occupational noise exposure when sound levels exceed those of the OSHA noise standard.

Head Protection

Falling objects are real dangers in many plants. Protective helmets, usually called hard hats, protect workers from head injuries. They also protect against water, grease, dirt, and sparks. Insulated hard hats offer additional protection. They are designed for workers exposed to electrical shock hazards.

In areas where less severe head injuries may occur, bump hats are worn. Bump hats offer less protection than hard hats. They may be worn in areas where protruding objects, such as pipes or machine parts, could cause injury.

❑ Hard hats are provided and worn where danger of falling objects exists.

❑ Hard hats are inspected periodically for damage to the shell and suspension system.

Hand & Arm Protection

Burns, cuts, pinches, and chemical or biohazard exposure are hazards that prompt hand and arm protection. Different types of gloves protect hands from specific hazards. Some gloves are designed to protect hands from sparks, flames, and flying particles. These are generated during welding, cutting, and other processes that involve a torch or other heat sources. Some protect against chemicals or electrical shock. Cut-resistant gloves protect against sharp tools or materials such as metal, glass, or abrasive materials.

Gloves cannot be worn around rotating equipment such as lathes, drill presses, and grinders. Here, barrier creams provide protection. Specially formulated barrier cream can protect hands from machine tool coolants, harsh chemicals, and other substances. Sleeves made of leather or other protective materials can be worn. They protect the arms from splashing liquids, extreme temperatures, or abrasive materials.

❑ Protective gloves, aprons, shields, or other means are required where workers could be cut or where there is reasonably anticipated exposure to corrosive liquids, chemicals, blood, or other potentially infectious materials.

Foot Protection

Disabling foot injuries are common in manufacturing jobs. Steel-toed shoes help protect feet in areas where heavy objects might fall on them. Slip-resistant soles provide traction and help prevent falls. Waterproof footwear is necessary at some work sites to keep feet dry and to prevent slipping.

❑ Appropriate foot protection is required where there is the risk of foot injuries from hot, corrosive, or poisonous substances, falling objects, or crushing or penetrating actions.

Lung Protection

Some airborne dust, fiber, particles, and gases put workers at risk for lung injury and illness. A dust mask, which is worn over the lower part of the face, filters out large particles. A respirator includes a face piece, hood, or helmet with a replaceable cartridge. It filters smaller particles and dangerous fumes from the air. A self-contained breathing apparatus (SCBA) is a respirator that supplies pure air to the user. The air is usually supplied from a tank of compressed air. Special training is required to use a SCBA. To ensure respirators are safe and effective, employers must provide those certified by NIOSH. Follow these guidelines for using respirators:

❑ Approved respirators are provided for regular or emergency use where needed.

❑ Use a respirator when work may expose you to fumes, oxygen deficiency, or particulates that a dust mask cannot filter.

(continued on next page)

    COPYRIGHT © GLENCOE/McGRAW-HILL

- Use a certified respirator with a cartridge designed for the particular contaminate. Certified respirators carry the term NIOSH on their label. The manufacturer's name and type of cartridge are also indicated on the label.

- Before beginning the work, do a fit test of the respirator. See **4-7, Conducting a Respirator Fit Test**, for further information about respirator fit tests.

Fall Protection

When workers are on scaffolds more than 10' above a lower level, they must be protected from falling. Fall protection consists of either guard-rails, safety nets, personal fall-arrest systems, or a combination of these. Personal-fall arrest systems stop a fall within a few feet of the worker's original position. Components include a full-body harness, a shock-absorbing lanyard, a rope grab, a lifeline, and a lifeline anchor. Body belts are not acceptable as part of this system.

- Lifelines are protected against being cut or abraded.

- Connectors are made of drop-forged, pressed, or formed steel or the equivalent, and are corrosion resistant.

- Each component is inspected before each use for any signs of wear, damage, or defects.

- If utilized in a fall, the system is then removed from service immediately to be inspected.

- Fall-arrest systems are attached above the worker's position to an anchorage or structural member that can support a minimum dead weight of 5,400 lbs.

- Fall-arrest systems are never attached to a guardrail or a hoist.

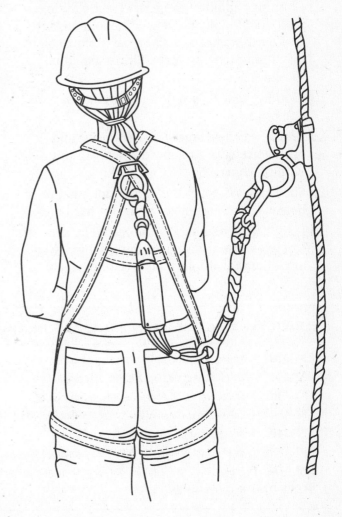

(continued on next page)

Clothing & Personal Wear

Protective clothing is needed for some types of work. For example, workers in steel mills wear protective clothing to shield them from heat. In areas where protective clothing is not necessary, workers should wear clothes that follow these guidelines.

❑ Materials must be safe in the given workplace environment. For example, flammable clothing is not acceptable for a torch operator.

❑ Wear short-sleeved shirts or roll up long sleeves to avoid getting caught in machinery.

❑ Pants and shirts should not have cuffs. Cuffs can collect sparks, dust, and other potentially hazardous materials.

❑ Clothing should fit properly. Tight clothing restricts motion and may cause an accident. Loose clothing can catch on parts of machines or fall into moving parts. Shirts should be tucked in.

❑ Clothing that is torn or has dangling parts is not worn.

❑ Rings, watches, bracelets, necklaces, and other jewelry are removed when working with machines.

❑ Shoes always cover the entire foot and are nonskid, well insulated, and with hard toes.

❑ Long hair is tied back or protected by a small cap when worker is operating machinery or any other hazardous tools or equipment.

Personal Cleanup

Some workers are exposed to hazardous materials that put their health at risk or could contaminate someone else. Many manufacturing environments provide areas for workers to wash or shower. They must shower before taking breaks and at the end of the workday.

Eyewash stations may also be provided. These small basins spray water upward to flush debris or chemicals from eyes.

❑ Adequate work procedures and protective clothing and equipment are provided and used when cleaning up spilled toxic or otherwise hazardous materials or liquids.

❑ Appropriate procedures are in place for disposing of or decontaminating PPE contaminated with, or reasonably anticipated to be contaminated with, blood or other potentially infectious materials.

❑ Eyewash facilities and a quick drench shower are available within the work area where workers are exposed to injurious corrosive materials. Where special equipment is needed for electrical workers, it is available.

MEETS THE FOLLOWING (MSSC) PRODUCTION STANDARDS

P1-07G	Knowledge of personal protective equipment (PPE) requirements, including safety shoes, goggles, and helmets.
P3-02E	Knowledge of clothing and PPE that should be worn to ensure safety.
P3-A3B	Select proper PPE for the job to prevent injuries.
P7-K3H	Safety and personal protective equipment is available, performs correctly, and has current certification.
P7-04D	Knowledge of PPE that should be worn.
P7-A15D	Read documentation on PPE needed when working on a tool.
P7-A17A	Explain HAZMAT requirements for equipment maintenance procedures (MSDS, PPE, OSHA).
CORE-01A	Knowledge of PPE.

COPYRIGHT © GLENCOE/McGRAW-HILL

Conducting a Respirator Fit Test

OSHA requires employers to conduct fit testing for all workers who will use a respirator. Before fit testing, the employer must provide training to workers, including the following steps.

❑ How to put on the respirator.

❑ How to position the respirator on the face.

❑ How to set strap tension.

❑ How to determine an acceptable fit.

The fit test must be performed while the worker is wearing any PPE that may be worn during actual respirator use. This is necessary to determine if the PPE will interfere with the fit. The worker should choose the respirator with the most comfortable fit and wear it for at least five minutes to assess comfort. Assessment of comfort should include a review of the following points:

❑ Position of the mask on the nose.

❑ Position of the mask on the face and cheeks.

❑ Room to wear eye protection.

❑ Room to talk.

The following points should be reviewed to determine the adequacy of the respirator fit:

❑ Chin properly placed.

❑ Adequate strap tension, not too tight.

❑ Fit across bridge of nose.

❑ Proper size to span distance from nose to chin.

❑ Tendency of respirator to slip.

❑ Self-observation in mirror to evaluate fit and position.

Once the comfort and adequacy of the respirator have been determined, the worker must conduct a user seal check. This must be done to make sure that the respirator seals properly and provides complete protection. If any part of the worker's apparel affects the seal, it must be altered or removed. The test cannot be conducted if there is any hair between the skin and the face piece sealing surface. This includes stubble beard growth, a beard, a mustache, and sideburns that cross the respirator sealing surface.

Additional tests are performed according to the type of respirator used and the hazard the worker might be exposed to. These tests can be found in OSHA regulation 1910.134, Appendix A, OSHA-Accepted Fit Test Protocols.

Mandatory User Seal Check Procedures

The individual who uses a tight-fitting respirator is to perform a user seal check to ensure that an adequate seal is achieved each time the respirator is put on. Either the positive and negative pressure checks listed by OSHA or the respirator manufacturer's recommended user seal check method shall be used. User seal checks are not substitutes for qualitative or quantitative fit tests. The respirator manufacturer's recommended procedures for performing a user seal check may be used instead of the positive and/or negative pressure check procedures provided that the employer demonstrates that the manufacturer's procedures are equally effective.

> **Refer to OSHA regulations (Regulation 1910.134, Appendix B-1) or the appropriate instruction manual for specifications and special procedures.**

Face-Piece Positive Pressure Check

For most respirators, this method of leak testing requires the wearer to first remove the exhalation valve cover before closing off the exhalation valve. The exhalation valve cover must then be carefully replaced after the test.

1. Close off the exhalation valve and exhale gently into the face piece.

2. The face fit is considered satisfactory if a slight positive pressure can be built up inside the face piece without any evidence of outward leakage of air at the seal.

(continued on next page)

Face-Piece Negative Pressure Check

The design of the inlet opening of some cartridges cannot be effectively covered with the palm of the hand. The test can be performed by covering the inlet opening of the cartridge with a thin latex or nitrile glove.

1. Close off the inlet opening of the canister or cartridge(s) by covering with the palm of the hand(s) or by replacing the filter seal(s).

2. Inhale gently so that the face piece collapses slightly and hold the breath for ten seconds.

3. If the face piece remains in its slightly collapsed condition and no inward leakage of air is detected, the tightness of the respirator is considered satisfactory.

MEETS THE FOLLOWING MSSC PRODUCTION STANDARDS	
P1-07G	Knowledge of personal protective equipment (PPE) requirements, including safety shoes, goggles, and helmets.
P3-02E	Knowledge of clothing and PPE that should be worn to ensure safety.
P3-A3B	Select proper PPE for the job to prevent injuries.
P7-K3H	Safety and personal protective equipment is available, performs correctly, and has current certification.
P7-04D	Knowledge of PPE that should be worn.
P7-A15D	Read documentation on PPE needed when working on a tool.
P7-A17A	Explain HAZMAT requirements for equipment maintenance procedures (MSDS, PPE, OSHA).
CORE 01A	Knowledge of PPE.

COPYRIGHT © GLENCOE/MCGRAW-HILL

Extinguishing Fires

Some employers prohibit employees from fighting fires in the workplace. Their emergency plans call for everyone to evacuate the site. Other employers, however, provide portable fire extinguishers for employee use. According to OSHA, if fire extinguishers are provided, the employer must also provide training to familiarize at least some employees with their use and with the hazards involved in fighting early stage fires. This training must be provided as soon as initial employment begins and at least annually from then on. Your employer must be in compliance with OSHA's requirements.

Training Others to Use Fire Extinguishers

If you are trained in fire extinguisher use, you might be asked to train others. You will need to develop skills in conducting this training.

❑ The three elements that must come together to start a fire, or to cause combustion, are:
- Oxygen (in the air and in some fuel materials).
- Heat (open flames, hot surfaces, sparks and arcs, friction, electrical energy, compression of gases).
- Fuel (gases, such as propane, acetylene, and carbon monoxide; liquids, such as gasoline, turpentine, and paint; and solids, such as wood, paper, plastic, and dust).

❑ Explain that there are four classes of fires. They are classified by the type of fuel that is feeding them. Each type of fire is best put out using an extinguisher designed for it. Using the wrong type of extinguisher may be dangerous.
- **Class A:** Ordinary combustibles, such as wood, rags, rubber, and plastics.
- **Class B:** Flammable liquids, such as gasoline, oil, grease, paint, and thinners.
- **Class C:** Electrical equipment, such as motors, switches, and electrical wiring.
- **Class D:** Combustible metals, such as iron, magnesium, sodium, and potassium.

❑ Before using any fire extinguisher, look at its label to be sure you have the right one. Pictorial symbols with blue backgrounds indicate which type of fire the extinguisher can be used for. If the symbol has a black background and a red line drawn through it, the extinguisher cannot be used for that class of fire.

❑ Check the label to make sure the extinguisher's testing is up-to-date. Extinguishers must be tested periodically to be sure their contents are still active. Dry chemical extinguishers, for example, should be checked for moisture. Maintenance requirements appear on the label.

❑ Make sure you have an unobstructed escape route behind you.

❑ Remember the PASS Sequence for fighting small fires:

P ull the pin to unlock the fire extinguisher's lever.

A im low, toward the base of the fire.

S queeze the lever to discharge the extinguishing agent.

S weep the hose back and forth across the fire.

(continued on next page)

Class of Fire		Type of Flammable Material	Type of Fire Extinguisher to Use	
Class A		Wood, paper, cloth, plastic	Class A	
			Class A:B	
Class B		Grease, oil, chemicals	Class A:B	
			Class A:B:C	
Class C		Electrical cords, switches, wiring	Class A:C	
			Class B:C	
Class D		Combustible switches, wiring, metals, iron	Class D	
Class K		Fires in cooking appliances involving combustible vegetable or animal oils and fats	Class K	

MEETS THE FOLLOWING (MSSC) PRODUCTION STANDARDS

P2-K4H	All appropriate safety equipment is present and in proper working order.
P2-O5C	Skill in conducting training on the use of safety equipment, such as fire extinguisher, eye-flush bottles, and first-aid kits.
P3-K2B	Emergency response complies with company and regulatory policies and procedures.
P7-O3C	Knowledge of certifications needed for regulatory compliance (i.e., cardiopulmonary resuscitation (CPR), fire extinguisher, and blood-borne pathogens).

COPYRIGHT © GLENCOE/McGRAW-HILL

Practicing Electrical Safety

Understanding electricity requires a basic knowledge of physics. Everything we see, touch, feel, or smell is composed of matter. All matter is made of atoms, and all atoms have an atomic structure.

The atomic structure of an atom includes three basic particles: protons, neutrons, and electrons. Protons and neutrons are found in the nucleus, at the center of the atom. Protons have a positive (+) electrical charge. Neutrons have no electrical charge. Electrons have a negative (–) electrical charge. Electrons circle the nucleus in different orbits, depending on their energy levels. These orbits are called shells. The electrons located in the outer shell are the important ones in the study of electricity. They contain free electrons that can move from one atom to another.

Conductors

An electrical conductor is a material that contains free electrons and easily conducts electrical current. The movement of free electrons through a conductor creates current flow, the movement of electrical energy through a conductor. It is important to know which materials are good conductors of electricity. These include copper (used in wiring), iron and steel, tin and lead, and gold and silver.

Insulators

Some materials do not have free electrons in their atomic structure. The lack of free electrons makes them poor electrical conductors. These materials are called insulators. They include plastic, rubber, and glass.

The Flow of Electricity

Voltage is a measurement of the pressure that causes electrical energy, or current, to flow. It refers to the electromotive force that moves electrons. Current flow is created when voltage moves electrons through a conductor. This flow is measured in amperes.

There must be a complete circuit before current can flow. Even though a voltage may be present, current cannot flow unless there is a return path to the current source. There are two types of current, direct current (DC) and alternating current (AC).

Direct current flows in a single direction. The direction of flow depends on the polarity of the applied voltage. Alternating current changes its direction of flow in a regular and predictable way. AC current moves back and forth in the circuit, reversing direction whenever the polarity of the applied voltage changes.

Resistance

Resistance is the opposition to current flow. The amount of resistance is measured in units called ohms. No material is a perfect conductor of electricity. Even copper, a very good conductor, has some resistance. The resistance of a conductor depends on the:

- Type of material from which it is made.
- Size of the material (for example, wire gauge).
- Length of the conductor.
- Temperature of the conductor. The lower the temperature, the less the resistance.

Working Safely

Electrical equipment and power lines are dangerous if not handled properly. They can cause severe shocks and burns. The body experiences electrical shock when it is part of the path through which electrical current flows. For example, touching a live wire and another wire at a different voltage can cause an electrical shock. Electrical shocks can cause severe injuries and death.

Many people believe it is the amount of voltage that determines the danger of an electrical shock. However, the real danger is the rate and amount of current moving through the human body and the path the current takes through the body. This means that a shock of 100 volts can be more deadly than one of 10,000 volts. The amount of current moving through the body depends on

(continued on next page)

resistance factors. These include the wetness of the skin, the length of contact, and whether the person is well grounded. A ground is a path for electricity to flow safely to the earth.

Arc blasts are a severe electrical hazard. They occur when powerful currents move through the air across a gap between conductors. Temperatures as high as 35,000°F [19,427°C] have been reached during an arc blast. This intense heat can cause burns. A high-voltage arc can also produce a serious pressure wave blast. This can cause a person to be thrown with extreme force.

Electrical safety procedures apply to any job involving electricity, not just those with exposure to high voltage or current. Even a very small current passing through the heart is extremely dangerous. Electricity demands respect and caution. To prevent shocks:

❏ Never operate electrical equipment without receiving proper instruction.

❏ Never open electrical enclosures unless authorized.

❏ Inspect and/or test all electrical equipment or lines before starting work.

❏ Use portable electrical tools and equipment only if they are grounded or double insulated.

❏ Do not use equipment if the electrical cord is frayed, cracked, or damaged in any other way.

❏ Never use equipment that has had the ground prong removed from the plug.

❏ Do not overload a circuit.

❏ Operate equipment only within its rated capacity.

❏ Keep hands, feet, and clothing dry, and use a dry board or a rubber mat when water or other moisture (including perspiration) is present.

❏ Change wet clothes and dry off completely before beginning electrical work.

❏ Stay away from broken electrical wires. Current may leak into the ground and put the worker in its path.

❏ Never use bare hands to determine if a circuit is live or touch live electrical parts with bare skin.

❏ Never use metal ladders, metal measuring tapes, or similar devices with metallic thread in areas where they could come in contact with electrical equipment or energized parts.

❏ Before replacing a breaker or fuse, determine why it needs replacement or notify a supervisor. There is a reason for a blown fuse or a tripped circuit breaker.

❏ Always replace a circuit breaker or fuse with a circuit breaker or fuse of the same capacity.

❏ Consider every electrical wire to be live until it is known to be dead.

❏ Know where the emergency shutdown switches are located.

❏ Ensure that all lockout/tagout procedures are followed.

❏ Make sure that adjustments or repairs are made only by authorized persons.

❏ Report as soon as possible any obvious hazard to life or property related to electrical equipment or lines.

❏ If there is an incident involving electricity, file an incident report, following the procedures in the company safety manual.

MEETS THE FOLLOWING PRODUCTION STANDARDS

P3-K1A	Potential hazards in the work area are identified, reported, monitored.
P3-O1I	Knowledge of how to inspect work area and report possible safety risks.
P3-O3F	Knowledge of OSHA and other health and safety requirements as applied to the workplace.
P3-A2B	Visually inspect work area for possible safety hazards.
P3-A8A	Work with co-workers to identify and report unsafe conditions.
P3-A17C	Knowledge of basic electrical systems to prevent electrocution.

   COPYRIGHT © GLENCOE/MCGRAW-HILL

Housekeeping Practices

In a manufacturing facility, it is vital to maintain tools, equipment, workstations, and the overall production environment. Practicing good housekeeping helps to:

- Maintain safety by keeping the environment clean and uncluttered and preventing safety hazards.
- Maintain efficiency by keeping tools, materials, parts, and other production items organized.
- Maintain the production schedule.

Housekeeping is an ongoing activity throughout the workday. All production workers should know and apply proper sanitation procedures in their work areas.

Aisles, walkways, stairways, storage areas, and work surfaces must be clear of tools, materials, debris, and clutter. Oil and grease on walkways and floors must be removed. Covered, fire-resistant containers reduce the dangers of oily rags and other flammable items. Approved and labeled containers with lids offer safe disposal or storage of hazardous wastes. Quick cleanup of dust, textile fibers, and other small particles that can catch fire or explode also lowers accident risk.

Slips and falls are one of the top three causes of injuries in the workplace. Clean and slip-resistant floors are the first defense against these accidents. In workplaces where workers work with liquids, employers must provide a properly working drainage system. Mats, platforms, and water-resistant, rubber-soled footwear also help workers avoid falls.

Exit Doors & Aisles

During an emergency, efficient evacuation depends upon workers knowing where all emergency exits are located. Exit areas must also be clearly marked and not blocked. Time is critical during emergencies, and the time it takes to unblock an exit could cost a life.

Aisles should be wide enough to allow workers to safely walk through the area. They also must remain clear, especially around machinery, and have sufficient clearance for walking and for motorized or mechanical handling equipment.

Storage & Organization

It's important to properly store tools and equipment, parts, materials, and other items used in production. Knowing where to find these items makes production run more efficiently. The old rule, a place for everything and everything in its place, really does help everyone. If an inventory of gears needed for a production run are not stored where they should be, valuable time can be wasted trying to find them. That can keep the team from meeting their production schedule. In addition, parts and materials must be stored safely to keep them from falling and injuring workers.

Racks, bins, shelves, and pallets should be provided for storage. These should be kept in good condition. The floor around them must be kept clear of obstructions.

Inspections

Housekeeping inspections can be scheduled or unscheduled. Make sure your work area will pass inspection at any time. Follow this checklist to keep a safe, efficient work area.

- ❏ Keep worksites clean, sanitary, and orderly.
- ❏ Keep walkways, stairways, storage areas, and work surfaces clear of tools, materials, debris, and clutter.
- ❏ Keep exit doors and aisles free of materials and equipment.
- ❏ Store combustible scrap, debris, and waste safely until it can be removed properly.
- ❏ Put oily rags and other highly flammable waste in approved containers.
- ❏ Remove oil, grease, and other liquid spills—including blood and other potentially infectious material—immediately according to proper procedures.

(continued on next page)

❑ Dispose of regulated waste (defined by OSHA's blood-borne pathogens standard) according to federal, state, and local regulations.

❑ Remove accumulations of combustible dust from elevated surfaces with a vacuum system that prevents the dust from going into suspension.

❑ Prevent metallic or conductive dust from entering or accumulating on or around electrical enclosures or equipment.

❑ Keep work surfaces dry and slip-resistant.

❑ Keep tools, materials, parts, and other necessary production items clean and organized.

MEETS THE FOLLOWING (MSSC) PRODUCTION STANDARDS

P2-K4	Perform all housekeeping to maintain production schedule.
P2-K4A	Tools are stored in proper location.
P2-K4E	Workstation is clean and clear of safety hazards.
P2-K4F	Scheduled housekeeping inspections are passed.
P2-K4G	Workstation is organized to maximize efficiency.
P2-A9B	Inspire production workers to maintain proper tooling storage in order to eliminate searching.
P7-K3G	Good housekeeping procedures are followed.

COPYRIGHT © GLENCOE/MCGRAW-HILL

Following HAZMAT Policies & Procedures

A hazardous material is a substance that poses a danger to human health or the environment. It could be something that poses a risk to human lungs when inhaled or catches fire upon contact with a spark. In manufacturing, hazardous materials (HAZMATs) are used to make products, operate machines, and clean equipment.

Types of HAZMATs

HAZMATs include chemical and biological materials that cause harm to your health or cause physical injury or death.

Hazardous Chemicals. These may be solids, liquids, gases, mists, dusts, fumes, or vapors. No matter what the form, they can be dangerous. Workers employed in chemical hazard areas must know the safety procedures for dealing with spills and other emergencies. Hazardous chemicals can cause injury and illness if they are inhaled, swallowed, or allowed to come in contact with the skin. The risk depends on the hazardous chemical and the amount and duration of exposure. Inhalation of certain chemicals for several hours or at high concentrations could result in illness or death.

Biological Hazards. Also called biohazards, these include bacteria, viruses, fungi, and other infectious agents that cause death, injury, or illness. Manufacturing work that deals with food and food processing in particular may expose workers to biological hazards. In these cases, workers should always wash their hands thoroughly with soap before eating. They should be careful about touching unprotected areas, such as the face.

OSHA Requirements

OSHA standard 1910.1200 states that its purpose "is to ensure that the hazards of all chemicals produced or imported are evaluated, and that information concerning their hazards is transmitted to employers and employees. This transmittal of information is to be accomplished by means of comprehensive hazard communication programs."

According to OSHA, hazard communication, or HAZCOM, must include:

- Identifying the material with a label.
- Providing a Material Safety Data Sheet (MSDS) that contains information about the health hazards and physical hazards of the material and procedures for protection from those hazards.
- Employee training.

The employer is responsible for establishing policies and procedures to meet OSHA's requirements. A key component of these procedures is employee training. The purpose of training is to ensure that every worker who handles hazardous materials has all the necessary knowledge and tools to identify hazards, use the proper safety equipment, and handle an emergency situation. According to the OSHA standard, "Employers shall provide employees with effective information and training on hazardous chemicals in their work area at the time of their initial assignment, and whenever a new physical or health hazard the employees have not previously been trained about is introduced into their work area." As an employee, you are responsible for using these policies and procedures to ensure your safety and that of your co-workers.

Materials Designated as HAZMATs

Hazardous materials are not restricted to the chemical or petroleum subindustries. Frontline workers in any subindustry may use HAZMATs at some time. Many manufacturing processes use solvents or chemicals that are hazardous. Cleaning and finishing products often present chemical hazards. Maintenance workers handle cleaning solvents, adhesives, and welding and soldering metals that may be hazardous. HAZMATs include materials that can cause health hazards and physical hazards.

(continued on next page)

Safe Use of HAZMATs

Follow these guidelines for learning how to safely use any type of HAZMAT you may work with or be exposed to.

❑ Know how HAZMATs are labeled. Some labels are standard throughout the chemical subindustry. Your workplace may have additional label information.

❑ Make sure that hazard labels remain intact on packages and containers. If a label becomes damaged, replace it so that other workers are aware of the hazard.

❑ Know where MSDS are stored. Your employer is required to have an MSDS accessible to you for every hazardous material on site.

❑ Be certain that you have received training on the handling of HAZMATs and use of a MSDS. If you have any questions about correct procedures or whether a material is hazardous, ask your supervisor.

HAZMAT Policies

Every company has a written HAZMAT policy that is designed to effectively communicate hazards and procedures for the particular hazardous materials that are used in its processes. By law, this policy must include the elements necessary for effective notification about and safe handling of hazardous materials. The following elements are included in a HAZMAT policy.

- Statement of the purpose of the policy.
- Written HAZCOM program.
- System for labeling all HAZMATs.
- MSDS readily accessible to employees for every hazardous material in the workplace.
- Employee training program that includes information on HAZCOM, identification of HAZMATs, and details on how to use MSDS.
- Employee training in protective measures and personal protective equipment (PPE) for handling HAZMATs.

APPLYING YOUR KNOWLEDGE

1. What are some potential dangers of working with biohazards?
2. What are the three elements that OSHA requires to be included in HAZCOM?
3. Name six elements an employer must include in a HAZMAT policy.

PRODUCTION CHALLENGE

You're starting a new job on the line at a facility that manufactures nail polish. You know that nitrocellulose and a number of solvents are used to produce the product. What information should your employer provide before you begin working?

MEETS THE FOLLOWING (MSSC) PRODUCTION STANDARDS

P1-07B	Knowledge of state and federal regulatory requirements (e.g., OSHA).
P3-02B	Knowledge of safety issues related to hazardous materials.
P3-03D	Knowledge of company safety standards for handling potential hazards.
P3-03F	Knowledge of OSHA and other health and safety requirements as applied to the workplace.
P5-03A	Knowledge of safety issues and practices including OSHA regulations to take or recommend action.
P7-01C	Knowledge of hazardous materials (HAZMAT) procedures information.
P7-04L	Knowledge of hazardous materials (HAZMAT) policies and procedures.
P7-A17A	Explain HAZMAT requirements for equipment maintenance procedures (MSDS, PPE, OSHA).
P8-08A	Knowledge of hazards in the workplace (i.e., spills, noise, air pollution) to ensure personal and fellow employee health and safety.
CORE-01C	Knowledge of HAZMAT procedures.

 COPYRIGHT © GLENCOE/McGRAW-HILL

Handling & Storing Chemicals

There are hazards involved in handling and storing certain chemicals. They include exposure to toxic materials, dangerous chemical reactions, and physical hazards of compressed gases. Workers must know the hazards and the appropriate procedures to safely handle and store chemicals. Every hazardous material has an identification label and a Material Safety Data Sheet (MSDS) that provides detailed information concerning hazards and handling of the material. In addition, most standard operating procedures include instructions and precautions for safe handling of materials that will be used. The key to safety when working with chemicals is to know the chemical with which you are working.

General Guidelines for Handling Chemicals

- Know your company's safety standards for handling potential hazards.
- Work around hazardous materials only if trained to do so.
- Read all labels and follow directions for handling hazardous chemicals.
- Never use an unfamiliar chemical until you have read the MSDS and know its hazards.
- Know which chemicals that you are using are hazardous and what type of hazard is involved, such as toxicity, flammability, or corrosiveness.
- Place hazardous chemicals in labeled containers only.
- Avoid skin contact by wearing protective clothing.
- Avoid breathing chemicals or getting them in your eyes. Goggles and a respirator may be required.
- Use a NIOSH-certified respirator when working with any airborne hazard.
- Use solvents only in well-ventilated spaces.
- Use explosion-proof power tools and nonsparking hand tools when using chemicals that can form an explosive gas mixture.
- Wash or shower thoroughly after handling chemicals.
- Remove, clean, or destroy all clothing that may have come into contact with hazardous chemicals.

Accidents often occur when containers holding hazardous chemicals are opened or moved. For example, some chemicals can explode on contact with each other or with compressed gases. To help prevent such problems, workers should:

- Avoid moving containers holding hazardous chemicals if possible.
- Determine the contents of all containers before moving them.
- Wear proper PPE whenever opening, moving, or transferring hazardous chemicals.
- Transfer contents of any ruptured or damaged container to a suitable container before moving.
- Avoid moving drums or containers that bulge, swell, or show crystalline material on the outside. These containers could be leaking or of poor quality, or they could have formed new and hazardous chemicals.
- Position containers to minimize contact with any sparks or heat sources when transporting hazardous materials by vehicle. Ensure that containers are properly secured during movement.
- Never use a torch or other type of tool that could ignite a fire to remove the lid of a container holding an unknown, possibly flammable, substance.
- Ensure that containers are grounded when moving a flammable liquid from one container to another. Ground wires prevent sparks from jumping between the containers and starting a fire.
- Stand clear of a container that is being opened, emptied, or moved by other workers.

(continued on next page)

Special Hazard Considerations

Some types of hazardous chemicals require special handling and storage procedures beyond the guidelines outlined above. These procedures will be included on the label and the MSDS. One particular concern is that chemicals can react with one another to make hazardous or toxic by-products. Workers must not handle these chemicals if they have not received the appropriate training.

Highly Reactive Chemicals

Some chemicals break down over time or when exposed to specific conditions such as heat, sunlight, shock, or friction. The products of this decomposition are often toxic or flammable and, in some cases, can cause fire or explosion. When working with highly reactive chemicals, you must pay particular attention to handling and storage precautions. This includes using specific types of tools and sealing containers to avoid contact with air or moisture. If you have any doubt about the proper procedures, consult the MSDS and/or notify your supervisor.

- Pyrophoric materials can ignite when exposed to air at temperatures below 130°F (54°C). These materials must be stored in airtight containers with an inert atmosphere or nonreactive liquid. In addition, they must be kept segregated from flammable materials. Examples are silane and phosphorus. Silane is used to manufacture fiberglass and to couple a bio-inert layer to titanium medical implants. Phosphorus is used in safety matches, pyrotechnics, and fertilizers and to protect metal surfaces from corrosion.

- Explosive materials can decompose rapidly, generating enough heat and gases to create a destructive explosion. Generally, these materials are stored in segregated areas, away from other chemicals, or in bunkers designed to withstand explosive forces. Examples are picric acid and concentrated hydrogen peroxide. Picric acid is used in the production of dyes and antiseptics. Concentrated hydrogen peroxide is used in organic and inorganic chemical processing, textile and pulp bleaching, and the production of antiseptic and cosmetics applications.

- Organic peroxides can decompose explosively. Organic peroxides and other chemicals identified as peroxide-forming compounds (which can be converted to peroxides over time) must be stored in areas that are segregated from flammable and other hazardous chemicals. Examples are benzoyl peroxide and isopropyl ether. Benzoyl peroxide is used as a bleaching agent for flour, fats, waxes, and oils, as a polymerization catalyst, and in pharmaceuticals. Isopropyl ether is a solvent for animal, vegetable, and mineral oils, fats, waxes, and some natural resins. It is also used in the manufacture of pharmaceuticals and fine chemicals.

Reactive Interactions

Some chemicals may be stable but have a tendency to react with other chemicals to form toxic or flammable materials. In this case, the chemical must be stored and handled in a way that will prevent reactive interactions.

- Water reactive chemicals react with water or moisture in the air, often producing flammable gases, and therefore, a fire hazard. When working with these materials, pay particular attention to precautions to prevent exposure to water. Do not store these materials near sprinklers. Examples are sodium, concentrated sulfuric acid, and acetic anhydride. Sodium is used in the production of a wide variety of industrially important compounds. Concentrated sulfuric acid is used to manufacture a wide variety of chemicals and materials including fertilizers, paints, and detergents. Acetic anhydride is used in the manufacture of cellulose acetate—the base for magnetic tape—and in the manufacture of textile fibers. It is also heated with salicylic acid to produce aspirin and in the manufacture of pigments, dyes, cellulose, and pesticides.

(continued on next page)

COPYRIGHT © GLENCOE/MCGRAW-HILL

- Oxidizers or oxidizing agents give up oxygen easily or oxidize other materials. It is important that these materials be stored away from flammable or combustible materials or materials labeled as reducing agents. Examples are chlorine and concentrated nitric acid. Chlorine is used widely to purify water, as a disinfectant and bleaching agent, and in the manufacture of many important compounds. Concentrated nitric acid is used in the production of fertilizers, rocket fuels and a wide variety of industrial metallurgical processes, including etching steel.

- Reducers or reducing agents react easily with oxidizers or acids and must be stored separately from them. Examples are magnesium and lithium aluminum hydride. Magnesium is used in structural alloys, pyrotechnics, and flash photography. Lithium aluminum hydride is used in the production of pharmaceuticals, perfumes, pesticides, and dyes.

Compressed Gases

Compressed gases, such as oxygen and acetylene, are used in many production processes, including welding and cutting. When stored under high pressure in metal cylinders, these gases can cause injury or damage. A broken valve could become a deadly missile. In addition, the contents could displace the oxygen in the room. Compressed gases also present potential reactive and health hazards. Special handling and storage procedures for compressed gases and pressurized vessels include the following:

- Store cylinders upright with the content label clearly visible.

- Store cylinders away from heat sources in a well-ventilated area.

- Secure cylinders in storage or in use from falling, either by chaining them to a wall or in a physical support system designed for cylinders.

- Separate cylinders based on their contents. Materials that are not compatible, such as oxygen and flammable gases (hydrogen, propane), should not be stored together. There should be at least 20 feet or a solid barrier wall between them.

- When moving cylinders, use a cart designed specifically for compressed gas cylinders.

- Never move a compressed gas cylinder that does not have its safety cap in place.

APPLYING YOUR KNOWLEDGE

1. What is the most important source of information about the hazards of a particular chemical?

2. Explain why pyrophoric materials must always be stored in an area that is segregated from materials that are labeled flammable or combustible.

3. Compressed gases must never be moved without a safety cap covering the valve. Why must this rule be followed even if the cylinder is only being moved a short distance?

4. Using the Internet or other resources, find a MSDS for epoxy paints. List the potential physical and health hazards associated with handling and storing these paints.

PRODUCTION CHALLENGE

Your work in a printing plant involves using the chemical methylene chloride in flexography processes. What steps should you take when handling this toxic chemical? How should the chemical be stored?

(continued on next page)

MEETS THE FOLLOWING ⬭MSSC⬭ PRODUCTION STANDARDS

P1-O7B	Knowledge of state and federal regulatory requirements (e.g., OSHA).
P2-K4B	Materials are kept in a safe manner.
P2-O4C	Knowledge of how to use and store hazardous materials and chemicals (e.g., compliance with MSDS).
P2-A17B	Knowledge of the chemical with which you are working.
P2-A17C	Understanding of chemicals so as to properly store dangerous materials and chemicals.
P3-O1L	Knowledge of procedures for handling hazardous materials.
P3-A17B	Understanding of potential chemical hazards.
P4-A17A	Knowledge of potential hazards of epoxy paints.
P4-A17B	Knowledge of chemical reactions.
P7-A17B	Apply knowledge of physics and chemistry to safety activities in the workplace.

COPYRIGHT © GLENCOE/MCGRAW-HILL

Preventing Spills

Facilities that handle hazardous materials are required by the Environmental Protection Agency (EPA) to have a Spill Prevention, Control, and Countermeasures (SPCC) Plan. The purpose of this plan is to protect the environment from the release of materials. If your company has an SPCC Plan, workers who handle materials that could be spilled will receive training on how to follow the plan.

An effective spill response and control plan should include:

- Spill and leak prevention measures.
- Spill response procedures.
- Spill cleanup procedures.
- Reporting procedures.
- Training in all of the above.

Even manufacturers who do not handle hazardous materials as part of the production process need to be concerned about spills. Vehicle fuels, cleaning solutions and solvents, and other common chemicals can become pollutants when they are released into the environment.

Preventing spills should be on everyone's mind in the workplace. Hazardous spilled materials can result in injury to workers, high cleanup costs, and possible long-term liability costs, such as paying to restore environmental damage. Many spills can be prevented by using good housekeeping techniques in the workplace and using care in handling materials.

An accumulation area is a special area set aside for storing hazardous materials and wastes. This area needs to be safely managed and should be restricted to people who are trained in handling hazardous materials.

General Spill Prevention

❑ Keep your work area neat and orderly—clutter increases the chance of spills.

❑ Make sure that there are appropriate waste containers for solid waste that could be contaminated, such as oily rags, or empty oil and solvent containers.

❑ Make sure containers have securely fitting lids that are kept closed.

❑ Store items on a hard, impermeable surface.

❑ If items are stored outside, make sure they are covered and secure.

❑ Do not store hazardous materials near doors or heavy traffic areas.

❑ Do not store hazardous materials near sanitary or storm drains.

❑ Do not place containers of liquid where they can easily be knocked over.

❑ Inspect containers regularly for rust, bulges, or leaks.

❑ Label containers and sites clearly with signs that read "Hazardous Waste" or "Hazardous Materials."

Handling Materials

❑ Use drain mats or plugs to protect sewers and drains if materials are handled near them.

❑ Use a drip pan, a large tub, or a basin as a secondary containment to catch any spills that do occur.

❑ Use funnels, drip guards, and other appropriate handling tools when transferring liquids.

❑ Use pumps and spigots for dispensing new solvents.

❑ Keep drums closed when not adding or removing materials.

❑ When moving drums of hazardous materials, use appropriate transport devices such as dollies.

❑ Make sure there is enough aisle space around drums to prevent puncturing from forklifts.

❑ Make sure that appropriate spill clean-up materials are readily accessible when handling materials that could spill. A cleanup kit should include appropriate absorbents and neutralizing materials and a plan for how to use them.

Safety Precautions to Prevent & Handle Spills

❑ Know where emergency contact numbers are posted before handling materials that could spill.

❑ Be aware of local fire codes and any additional fire, storm water, or other local and city ordinances that apply.

❑ Make sure that containers are compatible with their contents. For example, strong acids should never be stored in unlined metal containers.

❑ Ensure that overflow controls, alarms, relief valves, and leak detection devices are in proper working order.

❑ Check pumps, tubing, and pipes to be certain that they are not leaking.

❑ If a large spill or a spill of a hazardous material occurs, evacuate the area and notify your supervisor or a spill response team.

See **10-7: Handling Chemical Spills**, for additional spill response information.

MEETS THE FOLLOWING (MSSC) PRODUCTION STANDARD	
P7-01B	Knowledge of procedures to prevent or reduce emissions and spills.

COPYRIGHT © GLENCOE/McGRAW-HILL

Managing Hazardous Waste

The Environmental Protection Agency (EPA) and state environmental departments enforce the regulations that protect the environment and the public from harmful effects of hazardous waste. These regulations come from the Resource Conservation and Recovery Act (RCRA). There is a tracking system for hazardous waste. It is designed to ensure that hazardous waste is properly managed from the point of generation to the point of ultimate disposal. The EPA developed strict provisions for handlers of hazardous waste, including generators, transporters and disposers. All parties are given an EPA I.D. number and assume responsibility for their role in hazardous waste handling.

Identifying Hazardous Waste

The EPA's definition of a waste is "a material or chemical that has no intended use or reuse." Hazardous waste can be generated in a number of ways, including during the manufacturing process, equipment cleaning or maintenance, contaminated PPE, and contaminated materials used to clean up a spill. Hazardous waste regulations affect workers in many different parts of the manufacturing facility. Warehouse and shipping personnel handle the material when it is on site and as it is transported off site. Maintenance workers may come in contact with hazardous materials when working on process equipment. They may also generate hazardous waste when using cleaning or lubricating solvents.

Most hazardous wastes meet one of the following characteristics. If you are not sure whether waste meets one of the four characteristics, you should handle it as a hazardous waste. The hazardous waste can either be the chemicals described below or other materials contaminated by these chemicals.

- **Ignitable**—this includes materials that have a flashpoint below 140°F (60°C), including common solvents used in production and maintenance, such as alcohols, toluene, and acetone. Oxidizers and organic peroxides are also ignitable.
- **Corrosive**—these are liquids capable of corroding steel. They include strong acids, such as sulfuric acid and nitric acid, and strong bases, such as sodium hydroxide solutions.

- **Reactive**—these materials can react violently or create toxic fumes. They include reactive metals, such as sodium and potassium, and compounds, such as cyanides that can create toxic fumes if they come in contact with other chemicals.
- **Toxic**—a group of 40 materials are classified as hazardous due to their toxicity characteristic. Any detectable amount of these chemicals must be identified on the hazardous waste label.
- **Listed hazardous wastes**—the EPA has published a list of chemicals that are automatically determined to be hazardous waste. There are approximately 850 chemicals found on this list. If these materials are present above a certain minimum amount, which is defined for each chemical, they must be designated as hazardous waste.

Labeling Hazardous Waste

Under RCRA regulations, generators of hazardous waste must comply with specific requirements for labeling, marking, and placarding hazardous wastes for storage, shipment, and disposal.

Labeling Waste for Storage. If a facility generates specified levels of waste, the containers of waste must be marked with the words *Hazardous Waste* and the date that waste was first placed in the container. Hazardous waste must be disposed of within time limits that are defined by permits issued to the company, so careful records must be kept. These records must document the type of waste, the date it was generated, and the location of the waste container.

(continued on next page)

Labeling Waste for Shipment. Before hazardous wastes are transported off site, the containers must be labeled according to Department of Transportation (DOT) requirements. This ensures that people handling hazardous waste containers during transport can quickly identify both the waste and the potential hazards of the waste.

Placarding Vehicles and Containers. Each transport vehicle or freight container transporting hazardous waste must have placards, or warning notices, on each end and on each side. Placards must be readily visible and must be securely attached to the vehicle and meet specified standards for size, wording, and visibility.

Documenting Hazardous Waste

Hazardous waste must be documented from the time it leaves the generator facility where it was produced until it reaches the off-site waste management facility that will store, treat, or dispose of it. This tracking system allows the waste generator to verify that its waste has been properly delivered and that no waste has been lost or unaccounted for in the process.

Uniform Hazardous Waste Manifest. The key component of this system is the Uniform Hazardous Waste Manifest. The manifest is a paper document containing multiple copies of a single form. It is prepared by all generators who transport, or offer for transport, hazardous waste for off-site treatment, recycling, storage, or disposal. It provides the following information:

- Type and quantity of the waste being transported.
- Instructions for handling the waste.
- Signature lines for all parties involved in the disposal process.

The manifest is required by both the DOT and the EPA. Each party that handles the waste signs the manifest and retains a copy. This ensures accountability in the transportation and disposal processes. Once the waste reaches its destination, the receiving facility returns a signed copy of the manifest to the generator, confirming that the waste has been received by the designated facility. See the example manifest on p. 41.

Keeping Records. Every company that generates or handles hazardous waste is responsible for keeping complete records and filing reports with the appropriate federal and state agencies. Violating the record-keeping requirement can result in very large fines and other penalties.

Hazardous Waste Training

Any company or person who *knowingly* violates provisions of the RCRA is subject to criminal penalties. For this reason, hazardous waste documents should never be filled out by anyone who has not been trained on the requirements. Facilities that generate large quantities of hazardous wastes are required to provide annual training for employees. Workers who have not received this training should never sign documentation or manifests relating to hazardous waste.

Workers who handle hazardous wastes must receive training within six months of assignment. They may not work in unsupervised positions until the training has been completed. Employees must receive hazardous waste training if they are involved in managing or documentation of waste and if they might respond to, or be affected by, an emergency involving hazardous wastes. This includes frontline workers involved in any of these activities:

- Operating processes that generate waste.
- Moving or handling hazardous waste.
- Labeling or inspecting hazardous waste containers.
- Preparing Uniform Hazardous Waste Manifests or other documents.
- Emergency response.
- Equipment maintenance.
- Shipping.
- Anyone who supervises any of the employees listed above.

(continued on next page)

COPYRIGHT © GLENCOE/MCGRAW-HILL

Please print or type   (Form designed for use on elite (12 - pitch) typewriter)

Form Approved. OMB No. 2050 - 0039 Expires 9 - 30 - 91

UNIFORM HAZARDOUS WASTE MANIFEST	1 Generator's US EPA ID No.	Manifest Document No.	2. Page 1 of	Information in the shaded areas is not required by Federal law

3. Generator's Name and Mailing Address	A. State Manifest Document Number
	B. State Generator's ID
4. Generator's Phone ( )	

5. Transporter 1 Company Name	6. US EPA ID Number	C. State Transporter's ID
		D. Transporter's Phone

7. Transporter 2 Company Name	8. US EPA ID Number	E. State Transporter's ID
		F. Transporter's Phone

9. Designated Facility Name and Site Address	10. US EPA ID Number	G. State Facility's ID
		H. Facility's Phone

11. US DOT Description (Including Proper Shipping Name, Hazard Class, and ID Number)	12. Containers		13. Total Quantity	14. Unit Wt/Vol	I. Waste No.
	No.	Type			
a.					
b.					
c.					
d.					

G E N E R A T O R

J. Additional Descriptions for Materials Listed Above	K. Handling Codes for Wastes Listed Above

15. Special Handling Instructions and Additional Information

16. **GENERATOR'S CERTIFICATION:** I hereby declare that the contents of this consignment are fully and accurately described above by proper shipping name and are classified, packed, marked, and labeled, and are in all respects in proper condition for transport by highway according to applicable international and national government regulations.

If I am a large quantity generator, I certify that I have a program in place to reduce the volume and toxicity of waste generated to the degree I have determined to be economically practicable and that I have selected the practicable method of treatment, storage, or disposal currently available to me which minimizes the present and future threat to human health and the environment; **OR,** if I am a small quantity generator, I have made a good faith effort to minimize my waste generation and select the best waste management method that is available to me and that I can afford.

Printed/Typed Name	Signature	Month	Day	Year

T R A N S P O R T E R

17. Transporter 1 Acknowledgement of Receipt of Materials

Printed/Typed Name	Signature	Month	Day	Year

18. Transporter 2 Acknowledgement of Receipt of Materials

Printed/Typed Name	Signature	Month	Day	Year

F A C I L I T Y

19. Discrepancy Indication Space

20. Facility Owner or Operator: Certification of receipt of hazardous materials covered by this manifest except as noted in item 19.

Printed/Typed Name	Signature	Month	Day	Year

EPA Form 8700 - 22 (Rev. 9 - 88) Previous editions are obsolete.

(continued on next page)

APPLYING YOUR KNOWLEDGE

1. List two reasons why maintenance workers generally receive hazardous waste training.

2. What documentation is necessary in order to track a container holding items contaminated with a hazardous chemical when it is shipped?

3. Why is it essential that all hazardous waste documentation be prepared only by trained employees?

PRODUCTION CHALLENGE

Many processes used in the fabricated metal industry generate hazardous waste. These processes include machining unfinished metal workpieces for final assembly into a finished product. What recommendations would you make to your employer about reducing the amount of hazardous waste generated?

MEETS THE FOLLOWING (MSSC) PRODUCTION STANDARDS

P1-07B	Knowledge of state and federal regulatory requirements (e.g., OSHA).
P3-03D	Knowledge of company safety standards for handling potential hazards.
P3-03E	Knowledge of how to safely store, identify, and use hazardous materials and pressurized vessels.
P5-03A	Knowledge of safety issues and practices, including Occupational Safety and Health Administration (OSHA) regulations, to take or recommend action.
P7-01H	Knowledge of EPA required documentation for (a) disposal of hazardous wastes generated during maintenance or (b) transportation of contaminated items.

COPYRIGHT © GLENCOE/McGRAW-HILL

Planning & Conducting Tornado Drills

If your facility is located in an area subject to tornados, tornado drills are key to worker safety. The most important protection during a tornado is the ability to evacuate to a sheltered place quickly. An emergency plan and tornado drills ensure that, in a real emergency, an evacuation will be successful. Due to differences in layout and design, a unique plan is needed for every building and work site. This plan is frequently created by a safety team that includes workers familiar with various parts of the plant.

The plan should begin with determining the best evacuation route from any part of the structure to a designated shelter. Tornado shelters should be located in basements, interior rooms, or hallways, if possible. The plan should prioritize the procedures needed for training personnel in emergency response situations. Training should be delivered according to schedule.

Planning & Training

☐ Monitor changes in safety practice to ensure that training information is up-to-date.

☐ Determine tornado shelter areas for every building or work area. Interior hallways and stairwells are good places to shelter. Shelter areas should be:
- Enclosed.
- Windowless.
- In the center of the building, away from glass.

☐ Design an evacuation plan that includes at least two distinct routes to shelter for each area. If the primary route is blocked, the secondary route should be used. Evacuation routes must:
- Be as short and direct as possible.
- Be away from windows.
- Be through the interior of the building.
- Consider the needs of everyone who works in an area, including those with limited mobility.

☐ Post evacuation routes prominently in every work area.

☐ Make sure that the planned training will cover all topics and procedures needed to facilitate employee safety.

☐ Make it clear to all workers that failure to participate in the drill can lead to disciplinary action.

☐ Instruct all workers to use stairways, not elevators, for emergency evacuation. Elevators can become unusable during a tornado.

☐ Confirm that all employees, including new employees, are properly trained in emergency evacuation procedures. If necessary, schedule evacuation training prior to the drill.

Recognizing Signs of a Tornado

Training workers to recognize the signs of a tornado can be incorporated as part of the drill. This is especially useful for worksites where workers have no access to a weather radio. Besides the obviously visible tornado with a funnel, look for these signs:

- Strong, persistent rotation in the cloud base.
- Whirling dust or debris on the ground under a cloud base. Sometimes, tornadoes do not have funnels.
- Hail or heavy rain followed by either dead calm or a fast, intense wind shift.
- Day or night: loud, continuous roar or rumble that does not fade away quickly like thunder.
- Night: Small, bright blue-green to white flashes at ground level near a thunderstorm. These mean that power lines are being snapped by very strong winds.
- Night: Persistent lowering from the cloud base, illuminated by lightning.

(continued on next page)

Practicing the Tornado Drill

Tornado drills provide valuable information about strengths and weaknesses of the evacuation plan. Participation in tornado drills by all personnel is mandatory because practice improves the response during a real emergency when complications make evacuation difficult.

❏ Schedule drills so that all shifts and work areas participate on a regular basis.

❏ Position safety team members or other staff to observe the tornado drill. Pay particular attention to how well occupants of the building execute the evacuation. The effectiveness of the emergency plan depends on how well it has been communicated and practiced.

❏ Sound the tornado alarm and monitor the evacuation to the shelter areas. Tornado alarms must be independent of and emit a different tone than fire alarms.

❏ Make sure that all participants move to the shelter area in an orderly fashion.

❏ Make sure that all participants crouch down and cover their heads while in the shelter area.

❏ Confirm that all personnel have participated in the drill and evacuated according to the plan.

❏ Ensure that the training was understood. Give participants the opportunity to ask questions. They may have concerns or questions regarding safety issues or the training.

Evaluating the Tornado Drill

❏ Obtain feedback from observers and participants to determine whether there are any deficiencies in the plan or its execution.

❏ If problems are detected, modify the evacuation plan or schedule additional training as needed to correct the problem.

❏ According to company requirements, submit and file appropriate documentation to confirm the results of the drill and any follow-up actions that were taken. Report any issues or problems.

MEETS THE FOLLOWING (MSSC) PRODUCTION STANDARDS

P3-K2B	Emergency response complies with company and regulatory policies and procedures.
P3-K4D	Safety training is delivered regularly.
P3-A4A	Decide on the list of priorities necessary for training of personnel in emergency response situations.
P3-A5A	Organize safety drills to ensure worker safety.
P3-A5C	Plan the appropriate timing of emergency drills.
P3-A5D	Plan emergency drills to prepare for threats to health or safety.
P3-A6A	Communicate to the production supervisor that a safety issue exists and critical process must be stopped until a remedy is found.
P3-A6B	Interact with peers to share information on emergency drills/procedures.
P3-A6C	Interact with new employees on importance of safe work environment in order to make a positive impact.
P3-A6D	Give feedback to co-worker in order to communicate a safer way to perform an operation or task.
P3-A8B	Work with co-workers to identify and report unsafe conditions.
P5-K1A	Communication is sufficient to ensure that safety issues are understood and safety practices used.
P7-A11C	Constantly monitor changes in safety practices.

COPYRIGHT © GLENCOE/MCGRAW-HILL

Planning & Conducting Fire Drills

The most important protection during a fire is the ability to exit a burning building and move to a safe place quickly. An emergency plan and fire drills ensure that, in a real emergency, an evacuation will be successful. Due to differences in layout and design, a unique emergency plan is needed for every building and work site. This plan is frequently created by a safety team that includes workers familiar with various parts of the plant. The plan should begin with determining the best evacuation route from any part of the structure.

Planning & Training

❑ Design an evacuation plan that includes at least two distinct routes out of every area. If the primary route is blocked, the secondary route should be used. Evacuation routes must consider the needs of everyone who works in an area, including those with limited mobility.

❑ Post evacuation routes prominently in every work area.

❑ Make sure that the planned training will cover all topics and procedures needed to facilitate employee safety.

❑ Make it clear to all workers that failure to participate in fire drills can lead to disciplinary action.

❑ Instruct all workers to use stairways, not elevators, for emergency evacuation. Elevators can become unusable during a fire.

❑ Identify refuge areas to be used if evacuation routes are blocked. Refuge areas should include a closable door, a telephone, and a window.

❑ Designate an assembly area, well away from the building, where workers will gather and conduct a head count. In an emergency, the names and possible locations of missing personnel will be given to firefighters or the emergency team.

❑ Confirm that all employees, including new employees, are properly trained in emergency evacuation procedures. If necessary, schedule evacuation training prior to the drill.

Conducting the Fire Drill

Fire drills provide valuable information about strengths and weaknesses of the evacuation plan. Participation in fire drills by all personnel is mandatory because practice improves the response during a real fire when complications make evacuation difficult.

❑ Schedule drills so that all shifts and work areas participate on a regular basis.

❑ If the fire alarm is monitored by an outside agency, notify that agency of the drill to prevent an unnecessary response.

❑ Position safety team members or other staff to observe the fire drill. Pay particular attention to how well occupants of the building execute the evacuation. The effectiveness of the emergency plan depends on how well it has been communicated and practiced.

❑ Sound the evacuation alarm and monitor the evacuation.

❑ Meet at the designated assembly area and confirm that all personnel have participated in the drill according to the plan.

Evaluating the Fire Drill

❑ Obtain feedback from observers and participants to determine whether there are any deficiencies in the plan or its execution.

❑ If problems are detected, modify the evacuation plan or schedule additional training as needed to correct the problem.

❑ Submit and file appropriate documentation to confirm the results of the drill and any follow-up actions that were taken.

(continued on next page)

FIRE DRILL RECORD

Name of Facility:

Date of Fire Drill (mm/dd/yy)	Time of Day (am/pm)	Evacuation Time (min/sec)	Conducted By (staff name)	Problems Observed (action needed)
1.				
2.				
3.				
4.				
5.				
6.				
7.				
8.				
9.				
10.				

MEETS THE FOLLOWING (MSSC) PRODUCTION STANDARDS

P3-K2B	Emergency response complies with company and regulatory policies and procedures.
P3-A5A	Organize safety drills to ensure worker safety.
P3-A5C	Plan the appropriate timing of emergency drills.
P3-A5D	Plan emergency drills to prepare for threats to health or safety.
P3-A6A	Communicate to the production supervisor that a safety issue exists and critical process must be stopped until a remedy is found.
P3-A6B	Interact with peers to share information on emergency drills/procedures.
P3-A6C	Interact with new employees on importance of safe work environment in order to make a positive impact.
P3-A6D	Give feedback to co-worker in order to communicate a safer way to perform an operation or task.
P3-A8B	Work with all team members to conduct effective fire/safety/emergency drills.

COPYRIGHT © GLENCOE/McGRAW-HILL

Documenting Fire Drills

The following is general information regarding fire drill documentation. For specific local regulations, call the local Fire Marshal's office.

The person responsible for conducting and coordinating the building fire safety program shall immediately document improperly functioning fire safety equipment and arrange for repair and replacement as soon as possible.

Regulations

Specific regulations govern the completion and storage of fire drill documentation. Again, these are only general recommendations. The local Fire Marshal should be consulted to determine the type of documentation preferred.

❑ Documentation shall be available and kept in the appropriate office.

❑ Documentation should show that fire safety devices have been checked. Fire safety devices include fire extinguishers, smoke detectors, heat detectors, automatic fire alarm systems, emergency lighting systems, and sprinkler systems.

❑ Documentation must be kept for a specified period, usually 12 months to 24 months after the drill.

A Sample Fire Drill Report is included on p. 48 as an example.

Testing Fire Extinguishers

Following a fire drill is a convenient time to test and perform maintenance on fire extinguishers. Document the maintenance of the equipment and include a copy of the maintenance record with the documentation for the fire drill. Document the testing and maintenance of fire extinguishers on the following points:

❑ Fire extinguishers should undergo a monthly visual inspection. This is most likely the kind of maintenance that would be performed after a fire drill.

❑ Annual maintenance required for fire extinguishers can also be performed and documented after a fire drill.

❑ Hydrostatic testing and internal inspections should be performed and documented every 6 years.

MEETS THE FOLLOWING (MSSC) PRODUCTION STANDARDS

P3-K2	Perform emergency drills and participate in emergency response teams.
P3-K2A	Training and certification on relevant emergency and first aid procedures is complete and up to date.
P3-K2C	Emergency drills and incidents are documented promptly according to company and regulatory procedures.
P3-K3	Identify unsafe conditions and take corrective action.
P3-A14E	Document safety incident and training orientation.
P7-A11D	Know what is required for safety compliance and what is really needed to keep employees safe, to reduce cost and injury.

(continued on next page)

Sample Fire Drill Report

Date:		Time:	Location:			
Comprehensive Drill		**Silent Drill**	**Table Talk**		**Other**	

Instructions

Each department head, manager or designate is responsible for monitoring employee responses and assessing building features during every fire drill and at any time the fire alarm audible signal activates. Forward this completed form after each drill to (insert name of person and department).

Section 1	**Assessment of persons discovering / responding to fire**				

Describe fire drill scenario, fire incident or fire alarm occurrence:

	Yes	No		Yes	No
Simulated or Actual Activities					
Were people in immediate danger evacuated?			Zone of origin evacuated?		
Were doors closed and latched to confine the fire and reduce smoke spread?					
Was the fire alarm manually activated (if the scenario required this action)?					
Was the fire department called or switchboard notified as required by procedures?					
Was an attempt made to extinguish the fire?			Was attempt appropriate?		
Did sufficient staff respond and evacuate endangered occupants in an organized and timely manner?					
Was scene supervision appropriate?			Were instructions clear?		
Horizontal evacuation conducted?			Vertical Evac. Conducted?		

Comments/observations/recommendations on emergency responses:

Assessment of specialized Supervisory Staff responses	Yes	No
Was the fire department notified by phone promptly and correctly?		
Were verbal instructions correct and clearly stated over the voice communication system?		
Did designated staff respond correctly to provide fire department assistance and access?		

If "No" was answered for question(s) above, provide comments/observations/recommendations:

Section 2	**Did the following features operate properly in your area?**	Yes	No
A) fire alarm pull station (where applicable) and audible fire alarm devices			
B) voice communication system (voice messages were audible)			
C) self-closing doors closed and latched upon fire alarm system activation			
D) electro-magnetic locking devices released locked doors upon fire alarm system signal			
E) fire hose stations, fire extinguishers and/or sprinklers (where applicable)			
Section 3	**Did employees respond properly upon hearing the fire alarm signal and voice communication instructions?**	Yes	No
A) checked rooms and area for fire and closed doors immediately			
B) designated staff responded to the fire area to assist with evacuation			
C) hazardous equipment safely shut down where appropriate (i.e. oxygen, dryers)			
D) corridors were clear and unobstructed			

If "No" was answered for question(s) above, provide comments/observations/recommendations:

Print Name:	Signature:	Date:

COPYRIGHT © GLENCOE/MCGRAW-HILL

Understanding First Aid & CPR

Even when everyone in the workplace is careful, accidents can happen. If someone is injured, training in cardiopulmonary resuscitation (CPR) and first aid can reduce the possibility of permanent disability or death.

Always follow your employer's emergency response plan in case of accident or injury. This might mean summoning your supervisor or someone who has had training in first aid and CPR, or simply calling 911.

To be compliant with OSHA, if there is no clinic or hospital near the workplace, there must be an employee trained to give first aid. First-aid supplies or a first-aid kit must be readily available.

General First-Aid Actions

It is important for you to learn basic first aid, especially if you are working in an area where equipment and materials may be hazardous if not used properly. The following are the actions recommended by the American Red Cross.

❑ Rescue the victim and yourself.
❑ Restore or maintain breathing and heartbeat.
❑ Control heavy bleeding.
❑ Treat for poisoning.
❑ Prevent traumatic shock.
❑ Examine the victim carefully to evaluate injury.
❑ Seek medical help.
❑ Keep checking and assisting the victim until medical help arrives.

In addition, there are actions you can take if you're the first person on the scene of an injury, until trained first responders or medical personnel arrive. When dealing with an injured person, remember that the first rule is to *do no harm*. Careless or incorrect treatment can worsen the injury.

❑ Inform a supervisor.
❑ Call 911 or the nearest medical facility.
❑ Make sure the area is safe to enter.
❑ Keep the injured person lying down.
❑ If the person is not breathing, administer CPR, *if you are qualified.*

❑ Do not give liquids to someone who is unconscious.
❑ Treat for traumatic shock.
❑ Keep pressure on a wound to control bleeding, using latex or nonallergenic gloves.
❑ Keep broken bones from being moved.
❑ Keep heart attack victims quiet.
❑ For eye injuries, pad and bandage both eyes.
❑ Keep checking and assisting the victim until medical help arrives.

Cardiopulmonary Resuscitation (CPR)

CPR is a procedure designed to help a victim breathe and to restart the heart. It's a skill every production worker should have, especially those who work in even moderately hazardous areas. Most workplaces require several persons to be trained and certified in CPR.

> **CAUTION:** To learn CPR, you must be trained by qualified professionals. Injury can be caused if CPR is administered incorrectly. **Never** attempt CPR on a person who is breathing. Contact your local American Heart Association, American Red Cross, or hospital for information about classes for CPR and other first-aid training.

If you find an unresponsive adult, use the following steps to provide CPR:

1. Check responsiveness:
   - Tap the person's shoulder.
   - Shout, "Are you OK?"

2. If the person is unresponsive, shout for help or call 911 or the local emergency number. Follow your company's procedure.

(continued on next page)

3. Open the airway by:
   - Head tilt-chin lift.
   - If you suspect head or neck injury, use the jaw thrust.
   - Look in the mouth and remove any obvious obstructions.

4. Check for breathing. If the person is not breathing normally, provide 2 slow rescue breaths of 2 seconds each.
   - Be sure the chest rises with each breath.
   - If the chest does not rise, reopen the airway and try again.
   - Use a barrier device if available.

5. Check for signs of circulation, including normal breathing, coughing, or movement.
   - If signs of circulation are present but there is no breathing, provide rescue breathing (1 breath every 5 seconds).

6. If no signs of circulation are present, perform chest compressions on the lower half of the breastbone.
   - Compress at a rate of about 100 times per minute.
   - Do 15 compressions and then 2 rescue breaths (repeat).

7. Once you have provided CPR, recheck circulation.
   - After 4 cycles of 15 to 2, which is about 1 minute, recheck for signs of circulation.
   - If no breathing and no signs of circulation appear, resume CPR cycles, beginning with chest compressions.
   - If signs of circulation are present, but breathing is not present, give 1 rescue breath every 5 seconds.
   - Recheck for signs of circulation every few minutes or until arrival of rescue personnel.

MEETS THE FOLLOWING PRODUCTION STANDARDS

P3-K2B	Emergency response complies with company and regulatory policies and procedures.
P3-A11D	Acquire CPR and first-aid training.
P7-03C	Knowledge of certifications needed for regulatory compliance (i.e., CPR, fire extinguisher, and blood-borne pathogens)

COPYRIGHT © GLENCOE/McGRAW-HILL

Providing Basic First-Aid Measures

When you see that someone is injured, you might be able to provide basic first-aid measures. Whenever possible, a trained person should give first aid. However, if you are not trained but are the first on the scene of an accident, the following steps may be lifesaving.

Blood-Borne Pathogens. Blood-borne pathogens are microorganisms, such as viruses and bacteria, that are present in human blood and can cause disease in humans. They can be passed from one person to another through blood and other fluids. Employers should make available PPE, such as goggles and disposable gloves, so that workers assisting the injured will be protected from possible exposure. The person trained to give first aid should also be trained under OSHA's Blood-Borne Pathogen Standard. The purpose of this standard is to limit occupational exposure to blood and other potentially infectious materials. Any exposure could result in transmission of blood-borne pathogens, which could lead to illness or death.

Cuts & Scrapes

Minor cuts and scrapes usually will not require professional medical treatment. You can follow these steps:

1. Wear disposable gloves to avoid coming into contact with blood-borne pathogens.
2. Stop the bleeding with gentle pressure applied directly to the wound, using a clean cloth or bandage.
3. Clean the wound with clear water, not soap.
4. Apply an antibiotic to keep the surface moist and to discourage infection.
5. Cover the wound with a bandage to keep it clean.
6. Elevate the wound higher than the heart to slow bleeding.

Shock

A person can go into shock when the body fails to supply enough blood to the brain and other vital organs. Shock can result from injury, dehydration, sudden impact, or allergic reactions. Symptoms include nausea, weak or rapid pulse, dizziness, chills, shallow or labored breathing, pale skin, confusion, thirst, and unconsciousness. Shock can be fatal. In case of shock:

1. Elevate legs higher than the heart to keep blood flowing to the organs.
2. Keep the person warm, lying on the back if possible.
3. Seek medical assistance.

Falls

Determining the extent of injuries from a fall is difficult. The person may have broken bones or spinal or internal injuries. To avoid injuring the victim further:

1. Call 911.
2. Do not move the person at all, unless he or she would be in further danger.
3. Keep the person from moving by surrounding both the head and the body with folded clothes or blankets.
4. Treat any bleeding.
5. Keep the person warm and treat for shock.
6. Seek medical assistance.

(continued on next page)

Burns

For minor burns (first degree and second degree up to an area no larger than 2 or 3 inches):

1. Cool the burn under cold running water for 15 minutes or by immersing in cold water.
2. Cover loosely with a sterile gauze bandage.
3. Seek medical assistance if needed.

For major (third-degree) burns:

1. Call 911.
2. Do not remove burned clothing.
3. If necessary and *if you are trained to do so,* clear the person's airway and do CPR.
4. Cover the burn with a sterile bandage.
5. Treat for shock.

Electrical Shock

In case of electrical shock:

1. Call 911.
2. Examine but do not touch the person— he or she might still be in contact with the electrical source. Remember, you cannot help the person by becoming a victim yourself.
3. Turn off the source of electricity if possible and if it can be done safely. High power will require professionals.
4. If unable to disconnect power, break the contact if possible by using a *nonconductive* item, such as a wooden broom handle, to push the electrical source away from the victim.
5. Check breathing. If breathing has stopped, begin CPR immediately, *if you are qualified to do so.*
6. Treat for shock.
7. Treat for major burns.

MEETS THE FOLLOWING PRODUCTION STANDARDS

P3-K2B	Emergency response complies with company and regulatory policies and procedures.
P3-A11D	Acquire CPR and first-aid training.
P7-03C	Knowledge of certifications needed for regulatory compliance (i.e., CPR, fire extinguisher, and blood-borne pathogens).

COPYRIGHT © GLENCOE/MCGRAW-HILL

Writing Work Instructions

Written work instructions are a critical part of the day-to-day tasks in any manufacturing job. These instructions are sometimes known as work order instructions. These are the directions by which production jobs are performed. Work instructions should be carefully checked to ensure that they communicate the correct production information. This information communicates the customer needs to the production crew. Regardless of an employee's familiarity with a given task, work instructions are an invaluable reference. Work instructions may help in these instances.

- As training aids when training other workers.
- For a task that is only done occasionally.
- When even a small variation in technique may cause a nonconformance.
- To improve safety of hazardous tasks.

Examples of their use are:

- Instructions on how to complete a form.
- A listing of the settings and typical gauge readings on equipment.
- The key strokes necessary to complete a computer screen.
- Step-by-step instructions on processing a part through an operation.

Work instructions must be clearly written to minimize errors and maximize understanding. Work instructions should be tracked and documented as appropriate. In writing work instructions, it is important to:

- ❑ Use simple declarative sentences.
- ❑ Write in the present tense.
- ❑ Use graphics when appropriate.
- ❑ Add structure by using bullet points or an outline format.

Each manufacturing subindustry and company will have its own work instruction form. Familiarize yourself with the work instruction form used in your industry. A work instruction form presents information in a fairly standard format. The example form provided contains the following elements:

- **Heading.** This block contains the basic identification information for the document. It identifies the company, the title, number, issue date, and revision of the work instruction.
- **Purpose.** This block states the desired outcome that results from following the work instructions. The work instruction is written to produce this result. This statement of the purpose usually begins with "To . . . (do something) . . . "
- **Instructions.** The instructions constitute the list of actions necessary to accomplish the specified end. Generally the instructions are written in outline form. They also follow the sequence in which they are to be performed. This information should be carefully reviewed to determine if on-site adjustments are needed.
- **Approvals.** This block will identify the author of the document and the date it was written. It will also include a space for any necessary approval.

Providing work instructions for the setup of equipment and the handling of special materials will help ensure that everyone involved has a firm grasp of the procedures needed to perform the task correctly. Read, listen to, and understand all work instructions before beginning any assignment.

(continued on next page)

XYZ Manufacturing Company
Work Instruction Form

Title: *Work Instruction Form*   **Issue Date:** *MM/DD/YYYY*
Number: *001937*   **Revision:** *99*

Purpose: *To instruct the reader in the completion of a Work Instruction Form.*

Instructions:

1. **Use a standardized work instruction form.**
   a. *A standardized form simplifies data entry.*
   b. *A standardized form eliminates confusion for the reader that might be created when several different forms are used for the same purpose.*

2. **Write clearly.**
   a. *Be direct and to the point.*
   b. *Use simple declarative sentences (e.g.,Tighten the chuck.).*
   c. *Write in the present tense.*

3. **Include graphics.**
   a. *Where necessary, include pictures or diagrams to illustrate steps in the process.*

4. **Add structure.**
   a. *Add structure to the instructions by formatting them as an outline. This makes them easier to understand and helps eliminate miscommunication.*
   b. *Use bullets or similar formatting objects as appropriate.*

Written by: *The Supervisor*   **Date:** *MM/DD/YYYY*
Approvals: *The Plant Manager*

MEETS THE FOLLOWING (MSSC) PRODUCTION STANDARDS

P1-O4A	Skill in communicating work orders and customer needs to production crew to minimize errors and maximize understanding.
P1-O6A	Skill in interpreting work orders to meet customer needs.
P1-O6C	Skill in reviewing order sheets to determine if on-site adjustments are needed.
P1-A13A	Listen to and understand work instructions.
P1-A14E	Create written instructions for set up of production equipment.
P1-A14F	Develop written instructions for special material handling.
P1-A15D	Read instructions required to set up equipment and document process.
P2-A13A	Receive maintenance instructions and understand them.
P5-K2A	Communication reflects knowledge of material specifications.
P5-K4C	All parties are notified of production issues and problems in a timely way.
P5-K4G	Communications are tracked and documented, as appropriate.

   COPYRIGHT © GLENCOE/McGRAW-HILL

Completing a Change Order

A change order is a document that specifies and justifies a change to a contract. It communicates the change to different levels of management. It also explains why the change is needed. The change can sometimes involve a process. Typically, however, a change order changes the materials or the completion time. This in turn usually affects the total price.

A change order amends an existing purchase order. Reasons for amending an existing manufacturing purchase order include:

- To increase the value of a purchase order when additional goods are required or when the cost of such goods was originally underestimated.
- To decrease the value of a purchase order when the cost of the goods provided was originally overestimated or when the number of items purchased has been reduced.
- To amend the terms and conditions of delivery of goods cited under the existing purchase order.
- To extend or to shorten a vendor's, consultant's, or subcontractor's performance period.
- To cancel a purchase order.

To complete a change order, follow these guidelines.

- ❑ **Identify the change order.** Include the project name and number, contract name and number, change order date and number, and appropriate contract information.
- ❑ **Describe the change.** Provide a clear description of the change.
- ❑ **Justify the change.** Provide sufficient documentation to demonstrate that the change is needed.
- ❑ **Submit cost and pricing of the change.** This may prompt a review of the change. The greater the cost, the more detailed the review should be.
- ❑ **Obtain signatures of all necessary parties.** This ensures that everyone involved is aware of the change.
- ❑ **Submit the change order.** Send supporting documentation with the change order. Such documentation should acknowledge the changes in the terms of manufacture and delivery.

MEETS THE FOLLOWING (MSSC) PRODUCTION STANDARDS

P1-A7A	Change material or part to comply with new customer requirements.
P1-A10C	Communicate with different levels of management that a process needs to be changed and why.
P4-K6D	Adjustments are properly documented.
P5-A14B	Write change orders and document changes.
P8-O3G	Skill in compiling data and ensuring that changed processes or procedures have met new requirements.

(continued on next page)

REQUEST FOR CHANGE ORDER

SECTION A

Date: January 10, 2005   Dept #:   Req #:   PO #:

Vendor Name:

☐ Send to Vendor   ☐ Do Not Send; Internal Correction

Buyer Name: _____

CO #: _____

SECTION B

Requested Action: (Select at least one):   ☐ Add Item/Increase   ☐ Delete Item/Decrease

☐ Clear Open Commitment   ☐ Time Period Change   ☐ Cancel PO   ☐ Account Transfer

Reason(s) for Action: (Please attach any additional sheets)

SECTION C

Total Current Value of PO		Requested Change		Revised Total Value of PO	
Account	Amount	Account	Amount	Account	Amount
					$0.00
					$0.00
					$0.00
					$0.00
					$0.00
Net Total Value of PO before Change:	$0.00	Net Change:	$0.00	Net Total Value of PO after Change:	$0.00

SECTION D – Changes After Action: This Section is to be Completed by the Purchasing Office

Line Item #	Action Code	Quantity	Part #	Description	Commodity Code	Price	Extended Price

AUTHORIZATIONS:

Contact Person:	Authorized Signature:	Prior Approval (2nd Approval):
Telephone #:	Name/Title:	Name/Title:
Buyer	Purchasing Notes	Data Entry

COPYRIGHT © GLENCOE/McGRAW-HILL

Receiving Feedback

Feedback is communication from others that tells you how your actions are affecting them. The feedback received on the job from supervisors and co-workers can help you develop more effective work skills. Feedback allows people to check the effectiveness and appropriateness of their actions. It can also prompt them to modify or correct their actions.

Feedback should create a new awareness in the receiver. It is up to the receiver to learn from the feedback and decide how to act on the new knowledge. There are several situations in which feedback might be received. These include:

- Feedback from supervisors and co-workers regarding quality and safety.
- Feedback from co-workers regarding production needs.
- Feedback from machine operators regarding machine operations and safety.
- Feedback from team members in a continuous improvement process.

The effectiveness of feedback depends on the manner in which it is delivered and received. When receiving feedback, remember that it is important to distinguish what is perceived from what really happened. In receiving feedback, follow these guidelines:

❏ Listen carefully. You are seeking information and understanding.

❏ Be respectful. Do not interrupt to counter what is being said.

❏ Focus on the facts being presented. Anxiety often results from what we read into things and not from what people say. Your goal is to remain unaffected by any criticism directed at you. This will help you identify what needs to be done to improve the situation.

❏ Do not respond immediately. Unpleasant feedback can produce emotions of surprise, quickly followed by anger, rejection, or denial. Do not respond to the person giving the feedback until your emotions have settled down.

❏ Avoid becoming defensive. If you are criticized, calmly acknowledge the possibility that there may be some truth in the criticism. This allows you to be the final judge of what you are going to do.

❏ Reflect on the feedback. Analyze the feedback to determine what you think it means for you.

❏ Tell the person delivering the feedback that you'd like to discuss their feedback with them. This will show them that you care about their perceptions and needs. Set a time and a place for the discussion.

❏ Discuss the feedback. To ensure that you understand the feedback, follow these guidelines:

- Admit it when you do not understand what is being said. You will gain respect by asking for further explanation.
- Paraphrase. After listening carefully, restate the feedback in your own words.
- Summarize. Briefly restate the feedback information. This can confirm a shared understanding of what has been said or decided.
- Ask open-ended questions. Open-ended questions usually begin with words like What, Who, Where, When, and How. They are difficult to answer with a "yes" or "no" response. An example would be a question such as "What do you think about that?" Open-ended questions indicate your interest in learning more about the issues, ideas, and reasoning that are important to the person giving the feedback. Questions that ask "why" can put a person on the spot. Avoid asking such questions.

(continued on next page)

- Do not respond immediately. Wait for the other person to finish talking. Then pause rather than responding right away. Such a pause is not intended to embarrass the other person. By not speaking immediately, you let the other person know you are listening. You signal that you are interested.

❑ If you believe you were in error, admit it. By admitting the error, you can move forward. Do not be overly apologetic. Treat this as a learning experience.

❑ Agree on what changes on your part—if any—would be most helpful. If several changes are sought, ask which is the most important.

❑ Identify the changes you will commit to.

❑ Commit to undertaking specific actions by specific dates.

MEETS THE FOLLOWING (MSSC) PRODUCTION STANDARDS

P1-A13C	Receive feedback from supervisors.
P1-A13E	Listen to the needs of co-workers regarding production.
P3-A13C	Receive feedback from employees as it pertains to safety in a respectful and attentive way.
P4-A13A	Listen to the ideas of others in a non-judgmental manner to realize the greatest gain from the CI process.
P4-A13C	Receive feedback from supervisor on quality of work in an appropriate way.
P7-A6C	Communicate with operators regarding proper operations; also solicit feedback regarding machine operation.
P7-A13A	Listen to feedback from machine operators once a safety procedure is introduced.
P7-A13B	Listen to verbal communication of safety and OSHA regulations from supervisors (e.g. accidents, loss time, workers comp).
P7-A13C	Listen to and evaluate operator feedback for use in better safety procedures.

COPYRIGHT © GLENCOE/McGRAW-HILL

Providing Constructive Feedback

For feedback to be useful, it must be provided in such a way that the receiver does not feel threatened. Feedback is most effective when it is constructive and when it can be used to correct a problem. Providing constructive feedback can sometimes be helpful in meeting goals such as the following:

- Improving employee performance.
- Improving quality.
- Increasing production efficiency.
- Influencing others to accomplish production and quality goals.
- Influencing others to work quickly and safely, even in the face of hurried production.
- Suggesting continuous improvements.
- Suggesting corrective actions.
- Addressing on-the-job issues and concerns with team members, other shifts, and managers.
- Reporting to the original communicator issues that have been evaluated.

Feedback should be carefully planned and delivered. To help ensure that the feedback you provide is constructive, follow these guidelines.

☐ Be sensitive to the selection of feedback. The receiver needs to be ready to hear and deal with the feedback.

☐ Consider the needs of the receiver first. Focus the feedback on its value and usefulness to the receiver.

☐ Be sensitive to how much feedback to give. Present only the feedback that the person can use.

☐ Be sensitive to the timing of feedback. Make sure that feedback is given at an appropriate time.

☐ Provide the feedback in a calm manner. Do not raise your voice.

☐ Direct the feedback toward behavior that can be changed, not toward the person. Providing information about what a person does can help that person make choices about their behavior.

☐ Focus on actions, not motives. Feedback that relates to what, how, when, and where is based on observable events.

☐ Present the feedback as your observations. Do not make statements such as "People are upset when you are late." Instead, present such feedback by saying, "I am upset when you are late." With such a statement you express how you feel and identify the reason for your feelings. Simply saying, "You make me upset," does not identify a cause. Such a statement is more likely to lead to an argument. This will lead to less communication.

☐ Make your feedback specific rather than general and abstract. Feedback is generally more useful if it can be related to a specific time, place, and action. It is more useful to say, "I noticed that you broke in twice while I was speaking during the meeting" than, "You are always interrupting."

☐ Do not present opinions and judgments about the motive or intent of the person receiving the feedback. Such observations relate to interpretations and conclusions drawn from what was observed. Your opinions on motivation or intention will be based on your personal values. As such, your judgment will be subjective.

☐ Be objective. View the circumstances without personal feelings or interpretations. Present an objective description of what occurred, along with your reaction to the situation. By giving an objective description of what occurred and your reaction to it, you leave the receiver free to use the feedback as he or she sees appropriate.

(continued on next page)

COPYRIGHT © GLENCOE/MCGRAW-HILL

☐ Do not make assessments about the personality traits or personal qualities of the person receiving the feedback. Such assessments usually increase the person's defensiveness. For example, in addressing lateness, the statement "You arrived ten minutes late" is better than "You're irresponsible."

☐ Share information that will be helpful to the person receiving the feedback.

☐ Do not give advice. Giving advice influences a person's choice and lessens their full responsibility for their actions.

☐ Request what you'd like the person receiving the feedback to do differently. Let the person know exactly what you are asking them to do. By asking, you help them see what you think they could do to help the organization. They are free to accept, decline, or counter-propose.

☐ Check whether the person understands your feedback. One way of doing this is to ask the receiver of the feedback to say it in their own words. Then you can see if it corresponds with what you presented.

MEETS THE FOLLOWING (MSSC) PRODUCTION STANDARDS

P1-A9C	Influence other workers to accomplish production and quality goals.
P1-A9D	Influence others to work safely even in the face of hurried production.
P1-A12C	Provide feedback on process for quality improvement.
P2-K1G	Preventive maintenance schedule, documentation, equipment needs and outstanding repairs are communicated effectively from shift-to-shift, to team members, to managers, and to others as required.
P2-A12A	Suggest how a co-worker can improve work station efficiency.
P2-A12C	Provide feedback on machine condition in order to evaluate optimum usage.
P4-K3	Suggest continuous improvements.
P4-A9E	Influence line workers to take appropriate corrective actions as identified.
P5-K1B	On-the-job issues and concerns are discussed and quickly resolved.
P5-K1G	Issues are evaluated, tracked and reported back to original communicator.
P6-O2D	Skill in providing feedback to improve worker performance.
P7-A6E	Suggest safety improvements and interact with operators.

COPYRIGHT © GLENCOE/McGRAW-HILL

Using Social Skills

In the workplace, you must have the ability to establish good working relationships with people at different levels of the company. In other words, you must be able to interact positively not only with your peers but also with support staff and with management, including the president of the company. Using social skills is a way of developing those relationships. By doing so, you can develop more effective communication about work issues. Positive relationships with others facilitate effective work flow.

What Are Social Skills?

Social skills consist of communication between people that involves giving, receiving, and interpreting messages. We learn social skills at home, in school, in social situations, and on the job. How you use these skills can depend on:

- The situation you're in.
- Aspects of your personality.
- Past experiences.
- Your view of the other person.

Using social skills is verbal (words, sentences, grammar, and speech content) and nonverbal (eye contact, tone of voice, gestures, posture, facial expressions). Using social skills can be as simple as being friendly and courteous and showing respect for other people's ideas and feelings. It also includes acting ethically and having cross-cultural sensitivity.

The following checklist includes basic social skills you should use in the workplace.

- ❑ Act courteously and respectfully toward co-workers.
- ❑ Exhibit ethical practices in your daily work. This includes acting with honesty, integrity, respect, trust, responsibility, and good citizenship.
- ❑ Respect social and cultural differences in others.
- ❑ Show a positive attitude.
- ❑ Be a good team member.
- ❑ Follow instructions.
- ❑ Help and encourage others, especially in developing safe work habits and occupational and technical skills.
- ❑ Share your knowledge with others, especially new employees.
- ❑ Stay on task in your work.
- ❑ Communicate clearly when speaking and writing.
- ❑ Listen actively.
- ❑ Do not interrupt when others are speaking.
- ❑ Participate in resolving conflicts.

Facilitating Work Flow

The following checklist includes guidelines for using social skills in specific scenarios in the workplace.

- ❑ Establish rapport with maintenance workers so they will want to come to your area quickly when needed.
- ❑ Contact line workers in a friendly and enthusiastic manner to communicate safety and job-specific needs.
- ❑ Motivate employees through positive affirmations rather than intimidation and fear.
- ❑ Contact manufacturing leaders in a cooperative way in order to communicate process capability.
- ❑ Meet with operators to discuss quality issues in a nonthreatening manner to ensure that your message is heard.
- ❑ Initiate cooperation among work cell members to communicate production problems with one another.
- ❑ Inspire operators to help monitor the operations of new hires and temps.

(continued on next page)

Developing Social Skills

Although social skills are learned early in life, they can always be enhanced. New social skills can be learned too. Methods of learning these skills are often used together. They include:

- Direct instruction. The trainer teaches the importance of using different behaviors in social and work situations.

- Modeling. The trainer demonstrates, or models, positive social behavior on the job.

- Role-playing. After receiving instructions from a trainer, or role model, the role player acts out a real-life workplace situation using social skills.

MEETS THE FOLLOWING (MSSC) PRODUCTION STANDARDS

P2-A6A	Establish rapport with maintenance workers so they will want to come to your area quickly when needed.
P5-A6D	Contact line workers in a friendly and enthusiastic manner to communicate safety and job specific needs.
P5-A9B	Motivate employees through positive affirmations rather than intimidation and fear.
P6-K4E	Relationships with others facilitate effective work flow.
P7-A6F	Use social skills to sell the positives of good safety practices.
P8-A6A	Contact manufacturing leaders in a cooperative way in order to communicate process capability.
P8-A6D	Meet with operators to discuss quality issues in a nonthreatening manner to ensure that message is heard.
P8-A9A	Initiate cooperation among work cell members to communicate production problems with one another.
P8-A9C	Inspire operators to help monitor the operations of new hires and temps.

COPYRIGHT © GLENCOE/McGRAW-HILL

Operating LCD Projectors

The LCD (liquid crystal display) projector is used to display images from a source such as a computer or another electronic device onto a wall or screen for viewing. The projector is connected to the device by cables. Once connected, the projector projects an image from the source onto the surface the projector is facing. There are many different types of LCD projectors, each with its own set of operations. Refer to the user manual for your projector. The following steps outline the basic operation of the LCD projector.

1. Plug the projector power cord into the projector and then into a wall outlet.

2. Connect the projector to the source by means of video cables. Several types of cables can connect the source to the projector.

   ■ One type is a monitor cable that is used to connect a computer to the projector.

   ■ A second type is a video cable that is used to connect a device such as a DVD player to the projector.

   For more information about connecting the source to the projector, refer to the projector user manual.

3. If the presentation contains audio, connect the audio cables as well.

4. If you are connecting the projector to a computer, you may need to adjust the settings on the computer so that it can transfer the image to the projector. The video cards on some computers have more than one output connection. To display an image using the projector, you may need to tell the computer to activate the second output connection if it does not do so automatically.

5. After all of the appropriate cables have been connected, turn on the projector.

6. Turn on the source.

7. With the projector running, make sure that the proper input source is selected on the projector.

8. Position the projector so that the projected image can be seen clearly by everyone with a view that is minimally obstructed by the projector.

9. After the projector warms up and an image is displayed, focus the image.

10. Correct the color if necessary.

11. At this point, you can load a presentation onto the computer or other source device for transfer to the projector.

Making a Presentation

Giving an effective presentation involves more than simply operating the LCD projector. It is not complicated, but it does require some preplanning. The mark of any good presentation is that it is clear, easy to follow, and to the point. The requirements of each individual presentation will determine how it is produced, but the following are some general guidelines for making a presentation.

1. **Build an outline.** A presentation does not need to be long to be effective. In fact, it is better to do the opposite. The electronic slides should contain a basic outline of the material to be used as notes for speaking, rather than a comprehensive representation of the subject.

2. **Keep the slides simple.** Images, sound, and animation can liven up a presentation, but they also can be distracting. Use these elements sparingly. They should be used to emphasize the more important points you want to make. Omit them if they are not relevant to those points.

3. **Be succinct.** Be respectful of your audience's time. Give them the information they need without any unnecessary and distracting details. You are more likely to hold people's attention, and they are more likely to listen, if you are clear and concise in your presentation.

(continued on next page)

4. **Practice, practice, practice.** Go over the presentation beforehand and memorize the information you want to get across. Remember, the electronic slides are just an outline of the information so the audience can follow along and so you can refresh your memory. The bulk of the information should come from your speech. Avoid reading directly from the electronic slides. Talk to the audience, not to your notes. Make eye contact and speak so that everyone can hear you.

While you practice and when you give your presentation, keep these suggestions in mind.

❏ Speak up. Project your voice so that everyone in the room can hear you.

❏ Vary the tone and pitch of your voice. A presentation that is dry and given in a monotone voice will not hold an audience's attention for long. As much as you may care about the subject matter, it is hard to convince the audience to do the same if you sound bored.

❏ Speak with confidence. You are knowledgeable about your subject. Convey that to the audience and help generate credibility.

❏ Dress neatly. Khaki pants, a nice shirt, and a business dress/skirt are appropriate for most presentations. Your dress and appearance should convey a sense of professionalism.

❏ Make eye contact with audience members. Doing so helps hold their attention.

❏ Move about and use gestures to emphasize your speech. This helps hold your audience's attention.

❏ Take time to answer questions. Take the opportunity to clarify information that the audience might not have understood and reemphasize the major points of your presentation.

❏ Gather feedback. Ask your audience what they liked and/or disliked about your presentation. Apply this information to make future presentations better.

MEETS THE FOLLOWING PRODUCTION STANDARD

P5-A1A	Use overhead projectors and computers to train employees in the safe operation of equipment.

COPYRIGHT © GLENCOE/McGRAW-HILL

Practicing Effective Telephone Skills

The telephone is a powerful and cost-effective business tool. It can be used to make direct contact with customers on matters of quality and delivery times. When you use the telephone to talk to customers, correct telephone manners and etiquette will give a favorable image of your company to the caller. Poor telephone skills can damage relationships with internal and external customers.

Through experience, you'll develop your own telephone style. You'll also find customers and co-workers responding positively when you use the following guidelines as a model for your use of the telephone.

❑ Answer the phone quickly, within the first three rings.

❑ Answer the phone in a pleasant tone of voice.

❑ Identify yourself when you answer the phone. State your first and last name. If necessary, identify your company and the department.

❑ When you place calls, identify yourself when the other person answers the phone.

❑ Keep a notepad and pen by the phone at all times.

❑ Maintain a pleasant manner while on the phone.

❑ Never answer the phone while eating or drinking. Sounds associated with eating and drinking are amplified over the telephone.

❑ If you need to put a customer's call on Hold, explain why. Let the customer know how long it might take. Ask whether they would prefer to hold or to be called back. If they decide to hold, update them periodically if possible, rather than leaving them on Hold for a long time.

❑ Do not do other work while on the phone. The sounds will be heard by the person to whom you are talking. The listener will know that they do not have your full attention.

❑ Listen carefully to the person you are talking to. By focusing on them, rather than on what you want to say next, you will have more effective communication.

❑ Think about what you want to say before making an important call, so you can more effectively convey your ideas. If necessary, make brief notes identifying key ideas or problems that you wish to discuss.

❑ When calling someone with whom you need to have a fairly lengthy conversation, ask them if it is a good time to call.

❑ Make a telephone appointment for a call when you want to have a focused, longer conversation with someone who is normally busy.

❑ If others with you need to hear the call, ask permission before placing the customer on speaker phone.

❑ Make a telephone appointment for a conference call involving several people.

❑ If a customer leaves a message, return the call as soon as possible.

Evaluating Your Telephone Voice

Your voice reflects your personality over the telephone. It makes an immediate impression on the listener. You can control certain aspects of your voice to give the best impression. One way to learn how you sound to others is to make a recording of yourself while on the telephone. You can then evaluate the following attributes.

❑ **Quality.** The quality of your voice is its most individual characteristic. The way you "sound" is the feature that incorporates all of the features discussed below. It is the feature that makes your voice recognizable to others. It is important that the quality of your voice reflects understanding and professionalism. You do not want your voice to reflect a negative feeling such as anger.

(continued on next page)

❑ **Pitch.** Is your voice high or low? Do you speak in a monotone? Pitch varies in normal speech. These variations in pitch are known as inflection. Because use of inflection varies the tone of voice, it makes your voice more interesting. Inflection, however, can be overused. Also, keep in mind that when you are under emotional stress, the pitch of your voice will tend to rise. The pitch of your voice can suggest the state of your emotions to your listener.

❑ **Volume.** Is your voice loud or soft? Check the loudness of your voice. Speak loud enough to be heard, but not so loud that you sound forced.

❑ **Rate.** If you speak too slowly you may lose the attention of your listener. If you speak too rapidly, the listener may have difficulty understanding you. In either case, your message may not be clear.

❑ **Articulation and pronunciation.** Articulation is the way you pronounce words. It is important that you speak clearly. In business, it is important that you use the correct terms in the correct way. Faulty pronunciation and incorrect word usage can give your listener the impression that you are careless and lack knowledge.

Taking Messages

If you need to take a phone message for a co-worker, be sure to get all of the following information. Repeat the information to make sure it is correct.

❑ Name of the caller and company.

❑ Date and time of the call.

❑ Complete phone number.

❑ Any other information the caller provides, such as the subject of the call.

MEETS THE FOLLOWING (MSSC) PRODUCTION STANDARD

| P5-A1B | Use phone to communicate with customers on quality and delivery times. |

COPYRIGHT © GLENCOE/McGRAW-HILL

Communicating by E-Mail

E-mail is an invaluable tool for keeping in touch with customers, both external and internal. It can be used to inform concerned individuals about safety needs or conditions. It is especially useful when corrective actions are needed. For example, it can be used to communicate to all concerned parties the corrective actions that are being taken to meet specification limits. It can then be used to convey information to the necessary parties to ensure that the appropriate steps are taken to implement the corrective action.

Other advantages of e-mail include being able to:

- Communicate knowledge of customer and business needs in a timely and accurate manner to the correct parties.
- Track and document communication as appropriate.

With the increasing use of e-mail as a means of communication, it is important to know how to write e-mails properly. Your company may have specific policies regarding e-mail. The following are general guidelines.

❑ Know your recipient's communication preferences. Some people prefer the telephone to e-mail. Others prefer e-mail to the telephone.

❑ Ensure that the message is appropriate for e-mail. If it is more appropriate to deliver the message by telephone, in person, or by express or certified mail, do that.

❑ Do not use e-mail to resolve conflicts or to relay your disappointment or anger. Conflict resolution requires two-way communication. E-mail is one-way communication.

❑ Include a relevant Subject heading with the e-mail. If your message is just a few words, make your message the Subject.

❑ Do not identify the e-mail as Urgent or Priority unless it is.

❑ Organize the e-mail like a memo.

❑ Keep the e-mail as short as possible.

❑ Briefly describe who you are if the recipient does not know you.

❑ Do not shorten names (e.g., David to Dave) unless you know the shortened name is preferred by the recipient.

❑ Use lower-case and upper-case letters. Typing in all caps is considered SHOUTING.

❑ Use short paragraphs. Long paragraphs are hard to read in an e-mail.

❑ Maintain a businesslike tone.

❑ Proofread the e-mail. Check for spelling and grammatical errors.

❑ Do not send copies of e-mail unless the recipients need to be copied. Copies may make the recipients feel as if they have to do something with the information. It might also intimidate the primary recipient.

❑ Do not forward jokes.

❑ Do not send or receive at work e-mail that is unrelated to your job. The office PC belongs to the company. The company has the right to monitor its use.

❑ If you have not received a reply to an e-mail, do not assume you are being ignored. The e-mail may not have been received. Call to check.

❑ Treat each e-mail as confidential.

❑ When replying to an e-mail, follow the e-mail with your response. Keep the subject line as you received it.

❑ Do not send attached files unless you have a very good reason and only if the person is expecting them. Attached files can carry viruses. Some can take a long time to download, which can inconvenience the person who receives them.

(continued on next page)

❑ Consider the capabilities of your recipient. You may know that they do not have the version of the software you used to create the files you are attaching. You may also know that they have an earlier version and that your software is capable of saving a file in the earlier version. In that case, translate your files for them before attaching them to your e-mail.

❑ When sending an attachment, include information about the format of the attached file in your message. You might say, "Attached please find the latest draft of our recommendations in Word 97 format."

❑ For attachments, make use of naming conventions for certain types of files. For example, ending the name of a Word document with ".doc" and a WordPerfect document with ".wpd" will allow users of most Windows mail systems to automatically open the files in a suitable word processing program. These identifying name extensions are automatically supplied by applications on Windows.

❑ When attaching a picture or an image file, use a jpeg or bitmap format. If possible, save the file as a small to medium size file to save time in downloading.

❑ Respond to e-mails within two days.

MEETS THE FOLLOWING (MSSC) PRODUCTION STANDARDS

P5-K3D	Communication demonstrates knowledge of customer and business needs.
P5-K3F	Communication is made in a timely and accurate manner to the correct parties.
P5-K4G	Communications are tracked and documented, as appropriate.
P7-A1B, P8-A1C	Use e-mail to communicate the corrective actions taken to meet specification limits.
P8-A1D	Use fax or e-mail to convey information to necessary parties to ensure that appropriate steps are taken in the corrective action stages.

COPYRIGHT © GLENCOE/McGRAW-HILL

Operating Overhead Projectors

The typical overhead projector has two main parts: the light box and the projector head. The light box is just that, a box with a light in it. The top surface of the light box is a ground-glass screen that evenly diffuses light when the projector is turned on. The light box usually contains a cooling fan, which keeps the bulb from overheating and burning out.

The projector head consists of two ground-glass lenses positioned at a right angle to each other and connected by a mirror positioned at a 45° angle to the lenses, forming a triangle. The projector head is suspended above the glass surface of the light box on a moveable arm attached to the light box. One lens faces the glass surface of the light box. The other lens faces the surface on which the image is to be displayed. By placing or drawing an image on the ground-glass surface of the light box and turning on the projector, the image can be cast onto a large flat surface such as a wall.

The overhead projector is a simple and effective means of presenting information to a large group of people. It is a valuable instructional tool in training employees on the safe operation of equipment. Such training is even more effective when used with computer instruction.

Basic Steps

Refer to the user manual for the projector. If it is not available, refer to the following steps. They explain the basic operation of the typical overhead projector.

1. Plug the projector into a wall outlet.

2. Position the projector so that it faces the surface onto which the image is to be displayed.

3. Make sure that the projector does not obstruct the view of the audience.

4. Place or draw on the glass surface of the light box the image that is to be projected. Images for projectors usually come in two forms:

   - A transparent acetate sheet (or similar material) that carries a drawn, photocopied, or printed image.

   - A drawing directly on the glass surface of the light box. Note that drawings on the glass surface of the light box should be made with a marker intended specifically for that purpose. Other pens or markers can permanently mar the glass surface of the projector.

5. Turn on the projector. If the projected image is not in focus or of the appropriate size, adjust it.

6. To get the image to the proper size, move the projector closer to (smaller) or farther away from (larger) the wall or screen until the image is the desired size.

7. Focus the image by turning the focusing knob. The focusing knob is usually located on the arm by which the projector head is suspended. If the image will not focus when you turn the knob one way, turn it in the opposite direction.

8. Notice how the image changes size as the focus knob is turned. You may need to reposition the projector until the image is the right size. Repeat Steps 6 and 7 until the image is the desired size.

9. With the image properly sized and focused, you may need to reposition it so that it displays in the proper orientation on the projection surface. The image on the glass surface of the light box should be in the orientation that is desired on the wall or screen.

(continued on next page)

Additional Tips

Use the following tips when giving a presentation with an overhead projector.

❏ Use high-contrast images. If you are making transparencies with a photocopier or an inkjet printer, it is better to use higher contrast. These images are easier to see from a distance.

❏ Write directly on the transparencies, not on the ground-glass screen. Writing on transparencies makes the cleanup process easier. You need only remove the transparency rather than wipe down the screen to change topics. Writing on transparencies also helps protect the glass surface from damage.

❏ Prepare the presentation ahead of time on transparencies. It is tempting to write while you give a presentation. However, if you must give a long presentation or give more than one in a day, this writing can be time consuming and tedious. Prepare transparencies beforehand and write on them during the presentation only if it is necessary to supplement the existing information.

❏ Use color and graphics. Color and visuals, such as photos or drawings, add interest to a presentation. They help hold the audience's attention.

MEETS THE FOLLOWING PRODUCTION STANDARD

P5-A1A	Use overhead projectors and computers to train employees in the safe operation of equipment.

COPYRIGHT © GLENCOE/McGRAW-HILL

Creating PowerPoint® Presentations

Microsoft® PowerPoint is a powerful software program that allows the user to create electronic "slides" for presentation. It is relatively simple to use. A PowerPoint presentation can be effectively used in the manufacturing industry to provide a safety orientation. Using it, you can present accurate and cogent presentations to new hires and trainees on safety subjects. A basic presentation can be created quite easily. The following steps provide a basic tutorial on the program.

1. Open the PowerPoint program in one of the following ways:
   - If there is a desktop shortcut, double-click it to open PowerPoint.
   - Click the icon in the Start menu.
   - Click the icon in the Microsoft Office Toolbar (if the toolbar is active).
   - Use Windows Explorer to locate the program and click the icon to open PowerPoint.

2. PowerPoint usually opens with a blank "slide" on screen and the New Presentation Task Pane on the right-hand side of the screen. If there is no slide and the task pane is not present, bring the Task Pane into view by selecting Task Pane from the View menu.

3. In the upper left-hand corner of the task pane there are two arrows that allow the user to scroll through the task pane functions. Click the arrows to reveal the pane titled "New Presentation" if it is not already present.

4. With the Task Pane open, begin a new presentation in any of the following ways:
   - Click the "Blank Presentation" option. This brings up a title slide for the new presentation in the program's default format, which is black text on a white background.
   - Click the "From Design Template" option. This brings up a sampling of preformatted presentations on which text can be placed.
   - Click the "From AutoContent Wizard" option. This brings up a wizard that asks questions regarding the basic content of the presentation. The wizard uses this information to prepare a basic template into which the presentation text can be input.

5. Key in the text manually or copy the text from another document.

6. Place a title on each slide.

7. Precede each new idea on a slide by a bullet point.

8. Navigate between the text entry fields by using the mouse or pressing Ctrl+Enter on the keyboard.

9. When a slide is full, create a new slide. Do this in any of the following ways:
   - Click the New Slide icon in the toolbar.
   - Select New Slide in the Insert menu.
   - Press Ctrl+Enter when the cursor is in the last field on the slide.

10. When you have completed all of the slides, give the presentation a preliminary run-through to make certain that everything is set up properly. Select the first slide in the presentation.

11. With the first slide selected, click the View Show option in the Slide Show menu or press F5 on the keyboard.

12. Advance the slides by left-clicking with the mouse or pressing the space-bar.

A more complex presentation will require use of the advanced features in PowerPoint. For this, refer to dedicated reference material on the subject. Those who wish to use the more advanced features in PowerPoint should also use the Help feature. This feature is a good source of technical information. The PowerPoint documentation provides more detailed coverage of the information presented here.

(continued on next page)

Tips for Building PowerPoint Slides

When producing slides for a presentation, maintain a balance between simplicity and extravagance. On one hand, simplicity makes the presentation easy to follow, but too much simplicity will create a boring presentation. On the other hand, extravagance makes the presentation livelier and interesting, but too much can overwhelm the audience. They might miss the point of the presentation entirely. In general, slides should be just colorful enough to keep the audience's interest. Save pictures, graphics, and animation to emphasize the more important points.

❏ Choose a font that is easy to read. There are many options when it comes to choosing a font for your presentation. A cursive script font may look good on an invitation, but it can be hard to read during a presentation. Stick with a standard font such as Times New Roman, Arial, or Courier New.

❏ Choose a subtle background that enhances the text. Simplicity is critical in this situation. The background should not distract the viewer from reading the text. Black text on a white background is simple to the extreme. Although it works well for some presentations, using a lightly colored or patterned background to add some interest to the slides is preferable.

❏ Restrict yourself to a title and two to four bulleted points per slide. Remember, the slides serve as an outline for your audience and notes for you. Keep them simple and succinct.

❏ Use animation, sound, graphics, and photographs sparingly. Most presentations do not need them, but they can be used to add life to your slides and to help emphasize the major points of your presentation. Too much can be distracting.

MEETS THE FOLLOWING (MSSC) PRODUCTION STANDARDS

P3-A1D	Use PowerPoint presentations to conduct safety orientations.
P3-A12C	Present accurate and cogent presentations to new hires and trainees in safety subjects.
P7-A1A	Prepare and deliver PowerPoint presentations.

COPYRIGHT © GLENCOE/McGRAW-HILL

Writing Job Requirements

Job requirements documents will be read by cost estimators, technical specialists, and engineers. It is essential that the job requirements be understood not only by the writer, but also by the readers. Such documents are used to help co-workers understand the customer's job requirements. These documents are also used when meeting with customers regarding specific requirements. Everyone reading a job requirements document must interpret it in the same way.

To be easily understood, customer requirements as specified in job requirements must be well defined and written in plain language. Use of plain language saves time, effort, and money. Plain language requirements will vary from one set of requirements to the other, depending on the intended audience. In drafting customer's job requirements, follow these guidelines.

In all cases, job requirements should:

- Be definitive enough to protect the company's interests.

- Serve as a basis for customer evaluation and response.

- Provide a meaningful measure of performance so all interested parties within the company and the customer will know when the work is satisfactorily completed.

Use Active Verbs

Examples of active verbs include: calculate, create, design, build, develop, perform, and produce. These are all work words. For instance, the job requirements could require the manufacturer to "design and build the prototype."

❑ Avoid passive verbs that can lead to vague statements. Use a phrase such as "the manufacturer shall perform" rather than "it shall be performed." The latter phrase does not definitively state which party shall perform.

❑ Avoid the use of "should" and "may." The use of such words leaves the decision for action up to the manufacturer.

❑ Use "shall" when describing a provision that is binding on the manufacturer.

Be Concise

❑ Keep sentences short and to the point. Use no unnecessary words.

❑ Avoid redundancy. Redundancy, or unnecessary repetition, reduces the clarity of the message.

❑ If you need to amplify, modify, or make exceptions, make specific reference to the changes. Then describe them.

Use Understandable Terms

❑ Use simple well-understood words and phrases, except for needed technical terms.

❑ Do not use vague and ambiguous terms.

❑ Use descriptions that are careful and exact. This will help avoid misunderstandings.

❑ Ensure that each of the terms you use has only one universally understood meaning. Otherwise, include a definition that will provide a common basis for understanding between the customer and the manufacturer.

❑ Use acronyms and abbreviations only after spelling them out the first time they are used. If you use many acronyms and abbreviations, provide an appendix.

Be Specific

❑ Specify materials and processes.

❑ Do not use generalizations and inexact phrases. Phrases such as "etc.," "properly assembled," and "carefully performed" are examples of such unenforceable language.

❑ Avoid catch-all phrases such as "as directed" and "subject to approval."

❑ Avoid using "any," "either," or "and/or" unless the manufacturer is to be given a choice in what must be done.

(continued on next page)

Be Consistent

❏ Make sure that usage of terms is identical throughout the job requirements document when describing the same requirement. It can be confusing if a hole is referred to as an "orifice" and later called an "aperture."

❏ Make sure that usage of terms conforms to accepted standards.

Make Sure Requirements Are Traceable & Realistic

❏ Make sure that requirements are derived from the company's guidelines.

❏ Make sure the requirement can be achieved, produced, and maintained using the available resources and technology.

❏ Provide a realistic delivery schedule for contract performance and completion.

❏ Provide sufficient information for the manufacturer to establish its own milestones against which its progress can be measured.

❏ Provide information referring to data that will be used to evaluate whether the product is satisfying the requirement.

MEETS THE FOLLOWING PRODUCTION STANDARDS

P1-A6A	Meet with customers in a professional way to identify customer specific requirements.
P1-A9A	Help co-workers understand job requirements.
P8-O4G	Knowledge of customer requirements.

COPYRIGHT © GLENCOE/McGRAW-HILL

Writing Memos

The two most common types of business communication are letters and memos. A memo is a written note containing information, directions, or suggestions. It can be hard copy or an e-mail. Memos differ from business letters in these ways:

- Memos are usually a type of internal communication, used only within an organization.
- Memos are usually less formal than business letters.
- Memos are short and to the point. The writing style is direct.
- Memos are usually used for communicating nonsensitive information.
- Memos do not have a salutation or a closing.
- Memos have a specific format, with two sections.

Memo Format

The two sections of a memo are the heading and the body. The heading may include the word *Memo* or *Memorandum*. (Many companies use pre-printed memo paper that includes the heading and the company logo.) Beneath this are the following four lines:

DATE: (date when you write the memo)

TO: (person or people receiving the memo)

FROM: (person or people sending the memo)

RE: (short description of the subject)

The heading is followed by the body, which is the main text of the memo. The text should be well organized and well written. It should be concise and clear. Avoid using excessive technical jargon or formal language—use a more conversational style. Most memos should begin with a bottom-line statement, or a sentence that presents your main idea. In most cases, you do not need to provide background information because your audience will generally be familiar with the topic. In other words, you can get right to the point.

Memo Content

When you write a memo, be sure that it is concise and informative. To make a memo informative, you'll need to follow two important writing principles: preparation and organization.

Preparation. Determine your purpose in writing the memo. You might need to inform co-workers of a change in a customer's order or of an upcoming meeting. You should be able to state your purpose in a single, clearly written sentence. Knowing your audience is part of the preparation.

Organization. Present your ideas in a coherent, organized way. Begin your memo with a bottom-line statement, which expresses your main idea. Then follow up with material that supports and expands on that statement in logical order. If necessary, write an outline to organize your ideas. Do not omit any important ideas or details from the memo.

Write a Memo to Team Members

Memos are written by almost everyone in a company, from production workers to the CEO. If you need to communicate information to your team—for example, if customer needs change or if new equipment has been purchased—you might write a memo to let them know. Practice writing a memo to your team by following these guidelines:

- ❑ Identify your audience. As with any writing, consider what the audience already knows and what you will need to explain.
- ❑ Fill out the four lines of the heading.
- ❑ Make sure your subject line is concise and truly informs your audience what the memo is about.
- ❑ Do not use a salutation or closing.
- ❑ Use a bottom-line statement at the opening of the memo text.
- ❑ Keep the memo short by being concise and using simple, straightforward sentences.

(continued on next page)

- ❑ Use the first person (*I* or *we*) in a businesslike, but direct, conversational tone.
- ❑ Use separate paragraphs for new topics.
- ❑ Avoid using excessive technical jargon or formal language.
- ❑ If you have a list of items in your memo, use bullets to make reading easier. If you have several subtopics, use headings.
- ❑ Identify attachments (other relevant documents) at the end of the memo. If they become separated from the memo, your readers will know to ask for them.
- ❑ Proofread your work. Ask someone else to read it before you send it.

MEETS THE FOLLOWING (MSSC) PRODUCTION STANDARDS

P1-A14A	Write memos to communicate problems and changes to fellow team members.
P5-K1D	Communication demonstrates knowledge of customer and business needs.
P5-K1E	Communication is clear and relevant to the situation.
P5-K1F	Communication is made in a timely and accurate manner to the correct parties.
P5-A14D	Write reports and memos to staff regarding changes in requirements.
P6-A14E	Write team memos for education or informative purposes.

 COPYRIGHT © GLENCOE/McGRAW-HILL

Writing Goals

A project is a planned undertaking that has a beginning and an end. Those working on a project have a long-term goal of finishing the project. A long-term goal may take a year or more to achieve. To reach the long-term goal, they may have to attain several short-term goals. A short-term goal is one that can be attained in a relatively short period of time.

Team projects require that everyone on the team work toward that goal. By pursuing goals, you and your fellow team members can:

- Become more organized.
- Become more motivated.
- Focus on essential tasks.
- Influence other team members to pursue higher skill levels.
- Influence other team members to attain production and quality goals.
- Build decision-making skills.
- Develop problem-solving skills.

A team goal must be written in a succinct manner to motivate all those involved. Such a goal can help you identify key strategies. It focuses the energies of the team on meeting the team objectives. A well-written goal is a **SMART goal**, which is:

- **S**pecific.
- **M**easurable.
- **A**chievable.
- **R**elevant.
- **T**ime based.

By writing SMART goals, you will avoid writing goals that are unspecific, unmeasurable, and unattainable. In writing SMART goals, follow these guidelines:

- ❏ Meet with the team in a cooperative manner to establish goals and time frames.
- ❏ Specify the goal in writing. The goal should be specific, not vague or general. The goal should be a written action statement. It should be as concise as possible. Consider the first draft of the following goal and how it was improved when rewritten.

First draft: To increase production in the next year.

Rewritten: By December 31, 2007, to increase production of units of Product X on Line 7 at the Peoria, Illinois, facility by 17%.

- ❏ Be sure that the goal specifies a desired result that is aligned to customer and business needs.
- ❏ Encourage co-workers to share the common goal of satisfying the customer.
- ❏ Be sure that the goal specifies an expected amount of change.
- ❏ Be sure that the change specified in the goal is measurable.
- ❏ State the specific standards by which the amount of change will be measured. Examples of such measures are cost, percentage, quantity, and quality.
- ❏ Be sure that the goal cannot be reached too easily. However, the goal should not be so difficult that success is impossible. In writing production goals, a skill in identifying performance expectations is useful.
- ❏ In writing the goals, listen to the concerns of team members on their ability to meet and exceed certain goals. In setting the goal, their concerns need to be summarized and properly evaluated.
- ❏ Be sure the goal specifies the date by which the change is expected.
- ❏ Be sure that the goal is attainable within the designated time.
- ❏ Be sure that the goal specifies where the change is expected. This might be a geographical location, a population group, a production line, or a program.
- ❏ Break down a long-term goal into a series of short-term goals.

(continued on next page)

☐ In the case of a team effort, plan for each team member's contribution. In such a case, determine also how each team member's contribution will be measured.

☐ In the case of multiple goals, prioritize the individual goals.

☐ Document the goals and communicate them to all appropriate individuals.

MEETS THE FOLLOWING (MSSC) PRODUCTION STANDARDS

P1-A9C	Influence other workers to accomplish production and quality goals.
P1-A10B	Encourage co-workers to share a common goal—customer satisfaction.
P6-K2	Set team goals.
P6-K2A	Team goals are specific, measurable, and achievable.
P6-K2B	Team goals are aligned to customer and business needs.
P6-K2C	Team goals focus the team in order to meet team objectives.
P6-K2D	Team goals are documented and communicated to all parties.
P6-A6A	Meet with team in a cooperative manner to establish goals and timeframes.
P6-A9B	Influence others in the group to pursue higher skill levels.
P6-A13C	Listen to concerns of employees on ability to meet or exceed certain goals and summarize them for proper evaluation.
P6-A14D	Write team goals in succinct manner to motivate employees.
P8-O6A	Skill in identifying performance expectations to meet production goals.
CORE-A1P	**Leading others:** Motivate, inspire, and influence others toward effective individual or team work performance, goal attainment, and personal learning and development by serving as a mentor, coach, and role model and by providing feedback and recognition or rewards.

COPYRIGHT © GLENCOE/McGRAW-HILL

Solving Problems

Skills in problem solving are essential in manufacturing. The same procedure is used for solving both large and small problems. This procedure can be used to provide peer training on problem-solving techniques. Familiarity with this procedure will also allow more active participation when solving problems with a group or team. Following is the basic step-by-step procedure for problem solving.

1. **Identify and define the problem.**
   Before developing a strategy, a team needs to determine why the problem is occurring. Identifying the cause of the problem might involve interviewing people, analyzing data, and observing people at work.
   ❑ State the problem as clearly as possible.
   ❑ Strive for objectivity. Try to view the problem as if you were looking at it for the first time. If your view of the problem is influenced by your own interpretations and feelings, you may not arrive at the correct solution.
   ❑ Identify the circumstances that relate to the problem. Be specific. The circumstances might result from several different causes. For example, they might result from the physical environment, a mental attitude, an individual behavior, legal issues, or a lack of money.

2. **Generate possible solutions.**
   Brainstorming is a process in which all members of a group spontaneously present as many ideas as possible. Everyone calls out ideas, creating a "storm" of ideas. One member of the group writes the ideas on a flipchart. The only rule is that ideas cannot be evaluated until all of them have been presented.
   ❑ List all possible solutions.
   ❑ Do not worry about the quality of the possible solutions. A solution that at first seems unworkable may turn out to be the one that works.

3. **Evaluate possible solutions.**
   ❑ Determine which proposed possible solutions are unworkable or undesirable.
   ❑ Eliminate the unworkable and undesirable possible solutions.
   ❑ Evaluate the remaining possible solutions in terms of their advantages and disadvantages.
   ❑ Rank the remaining possible solutions in order of preference.

4. **Decide on a possible solution.**
   ❑ Decide on one possible solution.
   ❑ Specify who will implement the solution.
   ❑ Specify how the solution will be implemented.
   ❑ Specify when the solution will be implemented.

5. **Implement the possible solution.**
   ❑ Break down the activity into major steps.
   ❑ Create a schedule.
   ❑ Assign tasks to team members.

6. **Monitor progress.**
   ❑ Set interim goals.
   ❑ Meet regularly to make sure goals and deadlines are met.
   ❑ If some part of the solution is not working, the team should be willing to change its strategy.

(continued on next page)

7. **Evaluate the outcome.** Evaluation can provide some of the most important insights the team will gain. It can use these insights to improve future performance.

   ❑ Determine whether goals were met.

   ❑ Determine whether deadlines were met.

   ❑ If goals and deadlines were not met, determine what factors contributed to the problem.

   ❑ Determine what the team learned about their skills. Should they work on improving communication skills or building trust within the team?

   ❑ Evaluate the effectiveness of the solution.

   ❑ Decide whether the existing solution needs to be revised or whether a new solution is needed.

   ❑ If you need to identify and implement a new solution, return to Step 2 and repeat Steps 2–6.

MEETS THE FOLLOWING (MSSC) PRODUCTION STANDARDS

P5-A9A	Provide peer training on problem-solving techniques.
P6-K4F	Workers actively participate in meetings and problem-solving groups.
P8-A11D	Attend training in problem-solving techniques.

COPYRIGHT © GLENCOE/MCGRAW-HILL

Building Consensus

A consensus is a general agreement. Consensus building is a term used to describe a group participation decision-making process. Consensus building encourages participants in the process to present proposals or potential solutions for solving a problem. Each proposal is discussed to solicit points of common agreement or common disagreement.

Consensus building seeks to understand the issues and concerns of all participants in the process. It is not a competition. There is no majority rule and there are no winners or losers. All participants have equal input and veto power. In the end, all participants must agree (reach a consensus) on a common solution that is applied to solve the problem.

Consensus building is not a win-lose process. When a decision is reached by consensus, most team members accept the decision. They may not be completely satisfied, but they are committed to supporting it. Consensus decision making works because every team member has a voice. The team discusses advantages and disadvantages of the decision, and everyone's opinions are considered. Differing opinions can actually strengthen the decision.

Remember that consensus building does not mean that everyone agrees that the best decision has been made. It means that everyone can live with the decision that has been made.

Consensus building takes time. In fact, it usually takes longer than other kinds of decision making. Making decisions about important issues requires patience, time, and participation by everyone involved.

Benefits of Consensus Building

Many of the benefits of consensus building within a group relate to the building of trust and strength. Within a group, consensus building:

- Builds strong group involvement because everyone has a say in the final decision.
- Builds group trust by encouraging individuals with diverse interests to work together.
- Brings people with expanded areas of expertise and knowledge together to develop viable solutions.

- Encourages the decision-making group to be responsible for implementation and management of the solution.
- Reduces decision-making conflicts caused by majority rule or bargaining.
- Encourages empowering of individuals instead of overpowering by others.

Why use consensus building to find a solution? Why not take a hand or roll call vote and let numbers rule? Consensus building can be used:

- To build an environment that can overcome individual differences of thought and interest.
- To create a forum where all ideas are discussed and commonly acceptable solutions are agreed to.
- To encourage individuals to take ownership of a problem.
- To encourage individuals to become stakeholders and accept accountability and provide commitment.
- To eliminate any distrust that may result by not providing an arena for open discussion.

Building Consensus in Production

Building consensus among team members can prevent problems in the production environment. It can change behaviors in the following ways.

- Working with team members to determine the training needed to achieve measurable improvements in productivity and quality.
- Persuading others to ensure that equipment is operating correctly and good housekeeping is maintained.
- Facilitating agreement on machine maintenance schedules in order to minimize production impact.

(continued on next page)

- Resolving team member conflicts over workstation organization in order to create uniform setup.
- Explaining how to correct an unsafe condition without offending the affected workers.
- Reviewing potential or existing safety concerns.
- Building consensus by discussing potential actions needed to resolve potential or existing safety concerns.
- Facilitating agreement on safety procedures in order to ensure that the entire team follows the agreed-upon process.
- Creating consensus upon emergency procedures and specific people's responsibilities.
- Building consensus on what level of safety training is needed.
- Creating agreement that proper documentation of processes will help analyze areas that need improvement and provide insight on how to effect positive change.

Consensus-Building Guidelines

- Building consensus is not a competition. Everyone must trust each other and feel free to express their opinions.
- Everyone should contribute ideas.
- A facilitator should make sure that everyone is listening to and understanding each other and that the group stays on task.
- Disagreeing is fine, but do so in a respectful way.
- Try to separate the issue being discussed from personalities. Do not disagree simply because you do not like someone.
- Take time. There is no need to hurry.

The Consensus-Building Process

1. Identify the issue or problem to be resolved.
   - Stay on target and define objectives for the task.
2. Schedule a meeting.

3. Establish timeframes.
   - The decision process cannot be open-ended.
4. Identify the participants in the decision-making process.
5. Ensure that all participants understand the issues.
6. Assign a facilitator.
   - The facilitator acts as a neutral party, fully understands the issues, and keeps the discussion moving.
7. Determine what must be done to reach a solution.
8. Define what the outcome of the process should be.
9. Solicit proposals from each participant.
10. Make a list of all possible proposals.
11. Discuss all the proposals.
12. Solicit points of agreement or major disagreement on each proposal.
13. Eliminate proposals that generate major disagreement.
14. Revise the list and reduce the number of possible proposals.
15. Review each proposal.
16. Solicit points of major agreement or disagreement.
17. Retain proposals that are positive, with no major points of disagreement.
18. Reach agreement (consensus) on a proposal that satisfies the needs of the objective.
19. Assign individual accountability for each task involved in the proposal.
20. Reach agreement (consensus) on accountability.
21. Thoroughly document the results, especially areas of strong agreement and disagreement.

(continued on next page)

  COPYRIGHT © GLENCOE/MCGRAW-HILL

Concluding

❑ Do not have a preconceived idea for the outcome of the decision-making process.

❑ Determine how or if the decision satisfies the needs and concerns of the participants and that all participants are in agreement.

Evaluating

❑ Follow up on the decision and its outcome.

❑ Re-evaluate the outcome when unexpected events and situations change the anticipated results.

MEETS THE FOLLOWING MSSC PRODUCTION STANDARDS

P2-A10A	Work with team members to determine the training needed to achieve measurable improvements in productivity and quality.
P2-A10B	Persuade others to ensure equipment is operating correctly and good housekeeping is maintained.
P2-A10C	Facilitate agreement on machine maintenance schedule in order to minimize production impact.
P2-A10D	Resolve team member conflicts over work station organization in order to create uniform setup.
P2-A10E	Create agreement on the format of maintenance logs to ensure consistency.
P3-A10A	Explain how to correct an unsafe condition without offending the affected workers.
P3-A10B	Review potential or existing safety concerns and build consensus by discussing potential actions needed to resolve them.
P3-A10C	Facilitate agreement on safety procedures in order to assure entire team follows the agreed-upon process.
P3-A10D	Create consensus upon emergency procedures and specific people's responsibilities.
P3-A10E	Build consensus on what level of safety training is needed.
P4-A10E	Create agreement that proper documentation of processes will help analyze areas that need improvement and provide insight on how to effect positive change.

Resolving Conflicts

Conflict signals strong disagreement and results from differing viewpoints. The key to resolving conflict is managing it effectively. To do this, you need to learn the necessary skills. Conflict can occur in personal relationships. It can also occur within groups, as in meetings. Conflict management in all of these situations requires the same general skills. Here we'll discuss the management of conflict in a relationship with someone at work.

Effective conflict management requires practice. Note the underlying tone of respect in each of the following steps.

1. Try to determine if there is a problem between you and the other person.

2. If you think there is a problem, set up a private face-to-face meeting with the other person to discuss the problem.

3. Calmly and in a nonconfrontational manner, ask the person if there is a problem. If the answer is "No," inform the person that you think there is a problem. Explain what you think the problem is.

4. Agree to try to work peacefully together to find a solution to the problem.

5. Work together to establish ground rules. Agree that there is to be no interrupting or verbal abuse.

6. Invite the other person to describe the problem from his or her perspective.

7. Listen with an open mind. This means listening without immediately judging what the other person is saying. It also means listening to the other person as if you were not involved in the problem. This will help you arrive at a more objective view of the problem.

8. Do not interrupt. Pay close attention to what is being said.

9. When the other person has finished, check your understanding of the problem by summarizing it.

10. If necessary, ask questions to clarify the problem.

11. Try to understand the other person's perspective and feelings.

12. Let the other person know that you respect his or her opinion, even though it is different from yours.

13. Calmly state your position or opinion.

14. Ask the other person for feedback on your position.

15. Work together to identify the facts and issues you agree on.

16. Determine why different issues are important to each of you.

17. Identify options. Suggest possible solutions to the problem that would please both of you. Ask the other person to do the same.

18. Evaluate the options.

19. Compromise to reach a conclusion that is acceptable to both of you. It is important that each of you feel that your needs have been met.

MEETS THE FOLLOWING (MSSC) PRODUCTION STANDARDS	
P1-A10E	Resolve conflicts between two team members working together on a line.
P2-A10D	Resolve team member conflicts over work station organization in order to create uniform setup.
P4-A13B	Listen to the ideas of others with an open mind.

COPYRIGHT © GLENCOE/MCGRAW-HILL

Mentoring

How can new employees learn from the experiences of others? The answer is mentoring. Mentoring is a teaching method that enables an individual to share his or her personal experience and knowledge in an environment that can improve the professional skills and objectives of others.

In manufacturing, mentoring occurs when a more experienced worker (the mentor) gives significant career assistance to a less-experienced worker (the protégé). Mentoring relationships are particularly helpful for new employees.

Mentors are wise and trusted counselors for protégés. A mentor's knowledge, experience, tenacity, and skills offer the new employee guidance, advice, and training. However, while a mentor can steer a protégé in the right direction to reach his or her potential, protégés must still rely upon themselves to succeed.

Benefits of Mentoring

- Builds teamwork and instills a shared sense of commitment to the company's mission.
- Improves overall company performance.
- Encourages cross-training opportunities.
- Enhances and expands skill levels that improve business performance.
- Assists individuals in making critical career transitions.
- Produces confident and competent future leaders.

Qualifications of a Mentor

Why do you want to be a mentor? What can you offer your protégé? Be honest as you consider what influence, skills, and knowledge you can exhibit. Ask yourself the following questions to see if you qualify.

- Am I clear about my motives for helping my protégé? (If you're not sure yourself, the protégé will get mixed messages from you.)
- Do I show tact, diplomacy, and sensitivity in working with others who may belong to a different culture or are physically challenged?
- Do I demonstrate proficiency and initiative?

- Am I willing to share personal experiences relevant to the needs of the protégé?
- Am I able to identify my needs, expectations, and limits for the mentor-protégé relationship?
- Will I be able to look after my protégé's needs but consider my own as well?

The Mentoring Cycle

Scheduling

- ☐ Schedule regular meetings.
- ☐ Establish times for subsequent meetings.
- ☐ Come to each meeting prepared to discuss issues.

Guiding & Communicating

- ☐ Be honest about your experience and level of knowledge.
- ☐ Create a positive counseling relationship and a climate of open and positive communication.
- ☐ Be sensitive towards understanding the cultural and physical limitations of others.
- ☐ Empower the protégé to identify those areas where they feel they need help.
- ☐ Do not encourage or advise a protégé to engage in a task that you do not fully understand or would not do yourself.
- ☐ Understand differences in learning styles and modify your methods accordingly.
- ☐ Set goals so the protégé can gauge his or her level of confidence.
- ☐ Work to enhance and expand the skill levels of the protégé.
- ☐ Do not give up right away if your protégé resists your help at first. The protégé may not recognize the value of what you have to offer. Persistence—to a point—may help.
- ☐ Encourage the protégé to ask questions.

(continued on next page)

❑ Be open to all questions, no matter how trivial they may seem.

❑ Expose the protégé to work experiences not available in his or her immediate work environment.

❑ Share knowledge of organizational goals, policies, functions, communication channels, and training programs.

❑ Offer constructive criticism in a supportive way.

❑ Provide positive guidance based on your knowledge and personal business experiences.

❑ Assign "homework" if applicable.

Feedback & Problem Solving

❑ Encourage the protégé to provide feedback, both positive and negative.

❑ Provide positive feedback and personal encouragement. Mistakes can be a source of learning experiences.

❑ Identify specific work-related problems and be able to provide solutions.

❑ Help the protégé identify problems.

❑ Guide the protégé through the problem-solving process.

Concluding

❑ Do not have a preconceived plan for the outcome of the relationship.

❑ Be prepared for the relationship to end. A successful mentor-protégé cycle requires that the protégé moves on, and the relationship either ends or takes a different form.

❑ Identify mentors who have knowledge and personal experience in areas you may not.

Evaluating

❑ Evaluate your role as a mentor. Was your guidance beneficial to the protégé?

MEETS THE FOLLOWING (MSSC) PRODUCTION STANDARDS

P6-02A	Skill in training employees effectively, including on the job training.
P6-02C	Skill in matching training needs to business requirements.
P6-A9D	Mentor new employees.
P7-A7D	Understand differences in learning styles and modify methods accordingly.
CORE-A1P	**Leading others:** Motivate, inspire, and influence others toward effective individual or team work performance, goal attainment, and personal learning and development by serving as a mentor, coach, and role model and by providing feedback and recognition or rewards.

COPYRIGHT © GLENCOE/MCGRAW-HILL

Coaching

In manufacturing, a coach is someone who instructs another worker. Coaching includes explaining what needs to be done differently and discussing how to do it. A coaching session is usually a one-on-one meeting. Such meetings are helpful in maintaining effective and efficient work practices.

Certain circumstances within the manufacturing work environment can suggest the need for coaching. For example, coaching can be used to train a new employee. It can also be used to coach a co-worker on techniques that improve quality results to correct defects. It can be used to provide safety information. Coaching is behavior related or performance related.

Behavior-Related Coaching. A behavior-related coaching session can be used to discuss the employee's conduct before it reaches a level where more formal action would need to be taken.

Performance-Related Coaching. Performance-related coaching sessions can be used to discuss the employee's performance whenever it slips below the goals developed during the initial performance interview. Performance-related coaching sessions can also be used to build skills in an employee who needs to be introduced to a new process.

Benefits of Coaching

Your most valuable resource is time. Participating in a coaching session helps you maximize your use of time by maintaining a focus on one particular issue. By maintaining a focus, you help meet an objective. Regardless of your role in a coaching session, it can help you achieve your goals. Whether you are the coach or the person being coached, use each coaching session to realize the following important benefits.

- Sharpen your communication and interpersonal skills.
- Develop a mutual understanding of the other person's needs, perceptions, and goals.
- Build mutual trust.
- Clarify areas of difficulty.

- Allow conflict to be resolved at the lowest level.
- Build self-confidence and competence in your employees.
- Reassure, develop, and assist your employees.
- Release job tensions positively rather than negatively.

Guidelines for Coaching

Coaching can be used to reinforce effective performance and to correct faltering performance so it can be as effective as possible. If you are the coach:

❑ Provide the coaching at the appropriate time. Factors to be considered are:
  - Is this a time when the person will accept the feedback?
  - Will the person remember the specifics when it is time to perform?

❑ Focus on the behavior (both performance or conduct) and not on the person.

❑ Present a specific plan of action.

❑ Do not present a mixed message such as, "You did great, *but* . . . ."

❑ Coach in a way that allows the person to understand.

❑ Take into account the differences in learning styles and modify your methods accordingly.

❑ Suggestions should be data-driven. Their result should produce a measurable benefit to the customer, the company, and its employees.

Certain attitudes and behaviors are necessary if you are to derive the greatest benefit from a coaching session. This applies whether you are the coach or the person being coached. In a coaching session:

❑ Be open to feedback.

❑ Have a concern for helping others develop their skills.

(continued on next page)

❑ Be action oriented.
❑ Be organized and plan for the meeting.
❑ Use your interpersonal skills.

Feedback

Feedback is essential in any coaching session. Feedback that is correctly given and received becomes the basis for effective coaching. When feedback and coaching are used appropriately, performance and behaviors can be improved. If you are the coach, feedback from the person being coached can tell you how well that person understands your instruction. If you are the person being coached, feedback from the coach tells you how you are doing. The effectiveness of a coaching session will depend in part on the feedback involved. Much growth is a direct result of support in the form of feedback. The effectiveness of feedback depends on the manner in which it is delivered and received.

When giving feedback:

❑ Make sure that it is given at the earliest appropriate moment.
❑ Give it in an objective manner.
❑ Make sure that it is specific.
❑ Make sure that it focuses on behavior.
❑ Make sure that it describes the impact of the behavior on yourself and the situation.
❑ Use it to reward positive behaviors as well as to comment on negative behaviors.

When receiving feedback:

❑ Be open to feedback.
❑ Listen. Do not interrupt, justify, or explain.
❑ Keep an open mind.
❑ If necessary, ask for a specific behavioral example.
❑ Acknowledge the person giving you the feedback.

MEETS THE FOLLOWING (MSSC) PRODUCTION STANDARDS

P1-A9E	Coach and train a new employee.
P4-K3B	Suggestions communicate measurable and data-driven benefits to the company, its customers and employees.
P4-A9D	Coach a co-worker on techniques that improve quality results.
P7-A7D	Understand differences in learning styles and modify methods accordingly.
P7-A9E	Give proper training and coaching on how to safely use equipment.
P8-A9D	Coach operators in quality techniques to correct defects.
CORE-A1P	**Leading others:** Motivate, inspire, and influence others toward effective individual or team work performance, goal attainment, and personal learning and development by serving as a mentor, coach, and role model and by providing feedback and recognition or rewards.

COPYRIGHT © GLENCOE/MCGRAW-HILL

Training PIT Operators

Employers in general industry are required by the Occupational Safety and Health Administration (OSHA) to develop a training program specific to the type of truck that will be driven and working conditions that will be encountered. Employers must also evaluate the operator's performance in the workplace. The employer must certify that each operator has been trained and evaluated as required by the standard. One type of truck is a powered industrial truck (PIT).

All training and evaluation must be conducted by a person with the necessary knowledge, training, and experience to train operators and evaluate their competency. In order to teach others, this person must have solid knowledge of the information in the industrial truck training books, manuals, and service documents. Training activities should be planned to minimize production downtime.

If you are conducting an operator training program, keep in mind the following guidelines:

- Influence employees to attend equipment safety training.
- Promote and support on-site learning opportunities.
- Make suggestions and provide training in a courteous way.
- Interact with operators in a positive manner to ensure that proper safety methods are followed.

Training programs should be in place to instruct tool and equipment operators on proper procedures. The following operator training program relates to a PIT, commonly called a forklift. Note that this checklist outlines the main points of instruction.

1. **Introduction**
   - ❑ Ensure that the training techniques used are appropriate for the audience.
   - ❑ Explain that you are providing a training program based on the trainee's prior knowledge, the types of vehicles used in the workplace, and the hazards of the workplace.
   - ❑ Explain that the operators will be given the correct tools to do the job.
   - ❑ Use video, group discussion, and hands-on practice to allow each individual to build the knowledge and skills they will need to operate the truck correctly and safely.

2. **Truck Types, Features, and the Physics of Tipovers**
   - ❑ Give new operators a complete orientation to the equipment. Familiarize each operator with the basic types and functions of PITs.
   - ❑ Familiarize each operator with the information shown on a PIT's data plate. This plate contains the rated load capacity and maximum lift height, along with other information.
   - ❑ Explain the critical truck measurements that affect safety.
   - ❑ Explain the forces that cause tipovers.
   - ❑ Explain the industrial truck design considerations and safety ratings that help prevent tipovers, including the "stability triangle." This is the triangle formed by the wheels. The two front wheels are two points. The two back steering wheels connected on a central pivot are the third point. If the center of gravity moves outside of the stability triangle, the truck will tip over.

3. **Inspection**
   - ❑ Explain the purpose and importance of pre-operational checkouts.
   - ❑ Provide a basic understanding of areas covered during a pre-operational checkout.
   - ❑ Familiarize each operator with a checklist for pre-operational checkouts.

(continued on next page)

❑ Explain what should be done if a problem is found during the pre-operational checkout.

4. Driving

❑ Explain the elements of safe movement of a PIT.

❑ Explain the differences between an automobile and a PIT.

❑ Discuss the safety hazards associated with operating a PIT.

5. Load Handling

❑ Explain the elements of load lifting safety.

❑ Explain the safe operating procedures for raising loads in aisles.

❑ Explain the safe operating procedures for lowering loads in aisles.

6. Liquid Petroleum Gas (LPG) for Lift Trucks

❑ Discuss LPG and its properties.

❑ Present the procedures for safely refueling internal combustion vehicles.

❑ Describe the LPG tank components, including the service valve, surge valve, and relief valve.

❑ Discuss related safety issues.

7. Battery and Charging

❑ Explain the procedures for safely changing and charging batteries.

❑ Explain filling procedures.

❑ Explain maintenance procedures.

❑ Explain related safety issues.

8. Safety

❑ Communicate all important information regarding equipment safety clearly and effectively.

❑ Give the proper training and coaching on how to safely use the equipment.

❑ Review and reinforce the potential for serious injury.

❑ Review and reinforce the safety procedures to be followed in your facility.

9. Specific Hands-On Truck and Workplace Training

❑ Review features of the specific PITs to be operated.

❑ Review operating procedures of the specific PITs to be operated.

❑ Review safety concerns of the specific PITs to be operated.

❑ Review workplace conditions and safety concerns of areas where PITs will be operated.

❑ Guide operators in practicing the actual operation of specific PITs.

❑ Ensure that operators can demonstrate proficiency in performing the PIT operator duties specific to the trainee's position and workplace conditions.

10. Evaluation, Certification, and Documentation

❑ Evaluate operators by assessing their competency in performing the duties of the industrial truck operator.

❑ Provide certification of completion of the course to those who have received appropriate training, been evaluated, and demonstrated competency in performing the operator's duties. The certification must include
  - Name of the operator.
  - Date of training.
  - Date of evaluation.
  - Signature of the person performing the training or evaluation.

❑ Document the quality and effectiveness of the training according to company procedures.

❑ Forward the documentation according to company policy.

❑ Ask operators for evaluation and feedback.

❑ Use operator evaluation and feedback to improve training materials and methods.

(continued on next page)

COPYRIGHT © GLENCOE/McGRAW-HILL

MEETS THE FOLLOWING (MSSC) PRODUCTION STANDARDS

P1-A11D	Promote and support on-site learning opportunities.
P2-O5D	Knowledge of the certification/license requirements to operate specific equipment.
P2-A5B	Plan training activities to minimize production downtime.
P6-A11A	Learn how to train others (OJT) so that known skills are effectively communicated to others.
P6-A15A	Read and understand training books and manuals to disseminate information to team.
P7-K1A	New operators are given a complete orientation to the equipment.
P7-K1B	All important information regarding equipment safety is communicated clearly and effectively.
P7-K1E	Evaluations and feedback are utilized to improve training materials and methods.
P7-K1F	During training, trainee has the correct tools to do the job.
P7-K1G	Post-training evaluation indicates that workers can operate equipment safely.
P7-K1H	Training and facilitation techniques used are appropriate for the audience.
P7-K1I	Quality and effectiveness of training are documented appropriately.
P7-O3F	Knowledge of the tools and materials needed to operate equipment to train others.
P7-A2B	Orient new operators in the proper use of equipment and suggest process improvements.
P7-A5A	Plan safety-related training for equipment based on operator maintenance, installer experience.
P7-A5E	Develop plan to train all people who use equipment on safety procedures and practices.
P7-A6A	Interact with the user in positive manner to assure that proper safety methods are followed.
P7-A6B	Make suggestions or provide training in a courteous way.
P7-A6D, P7-A12D	Train others to use equipment safely.
P7-A8E	Share knowledge and experience with team members to increase the knowledge of entire team.
P7-A9D	Influence employees to attend equipment safety training.
P7-A9E	Give proper training and coaching on how to safely use equipment.
P7-A12B	Give presentation on proper use and safety of tools and equipment to co-workers.
P7-A15B	Read equipment operation and service documents.
P7-A15E	Read training materials, safety rules, and equipment operating procedures.
P7-A17B	Apply knowledge of physics and chemistry to safety activities in the workplace.

Documenting Training

Training employees in the proper performance of a job is time and money well spent. Those who are new on the job have a higher rate of accidents and injuries than more experienced workers. A knowledge of specific job hazards and proper work practices can help reduce on-the-job accidents and injuries. Such knowledge comes from proper health and safety training.

Many Occupational Safety and Health Administration (OSHA) standards require the employer to train employees in the safety and health aspects of their jobs. Other OSHA standards make it the employer's responsibility to limit certain job assignments to employees who are "certified," "competent," or "qualified." Such employees will have had special training, in or out of the workplace.

OSHA has developed voluntary training guidelines. These are designed to assist employers in providing the safety and health information and instruction needed for their employees to work at minimal risk to themselves, to fellow employees, and to the public. The following guidelines include the essential steps in the OSHA Training Guidelines.

1. **Determine the need for training.**
   - Problems that can be addressed effectively by training include those due to lack of knowledge of a work process, unfamiliarity with equipment, or incorrect execution of a task.
   - Training is less effective for problems arising from an employee's lack of motivation or lack of attention to the job.

2. **Identify training needs.**
   - ❑ Identify what the employee is expected to do.
   - ❑ Conduct a job analysis that pinpoints what an employee needs to know to perform a job.
   - ❑ Identify in what ways, if any, the employee's performance is deficient.
   - ❑ Observe employees at their workstation while they are performing their job tasks.

Training needs might be met by revising an existing training program rather than developing a new one. Training is most effective when related to the goals of the safety and health program. Ideally, such training should be provided before problems or accidents occur. Training should be repeated if an accident or near-miss incident occurs. The company's accident and injury records can help identify how accidents occur and what can be done to prevent them from recurring. Ask the employees to identify "at risk" activities. Focus on those issues when developing your training programs.

3. **Identify training goals.**

   Each goal must be SMART. A SMART goal is a goal that is
   - **S**pecific
   - **M**easurable
   - **A**chievable
   - **R**elevant
   - **T**ime based

   Each goal should specify as precisely as possible what the employee should do. Progress should be measurable. The goal should specify the conditions under which the individual will demonstrate competence and define what constitutes acceptable performance. The goal should be relevant to the employee's activities. A timeframe should be established for the employee to meet the goals.

4. **Develop training activities.**
   - Training activities should simulate the actual job as closely as possible.
   - Training activities should enable employees to demonstrate that they have acquired the desired skills and knowledge.

(continued on next page)

COPYRIGHT © GLENCOE/McGRAW-HILL

5. Conduct training.

- ☐ Make training relevant to the equipment, tools, materials, and processes at the workstation.
- ☐ Make sure that employees understand why a task must be performed in a certain way.
- ☐ Implement those training practices that will be most effective. For example, many training programs are now on-line.

6. Evaluate training effectiveness.

- ☐ Measure the improvement in employee performance.
- ☐ Measure the success of the overall training program by measuring the reduction in injury or accident rates.

7. Improve training if necessary.

- ☐ If the evaluation reveals that the training did not give employees the expected level of knowledge and skill, revise the training program.

8. Document training.

- ☐ Document training correctly according to company policy.
- ☐ The employee being trained should be able to demonstrate competence in the performance of the task before the supervisor completes the training documentation form.
- ☐ Update training records as needed.
- ☐ Make sure that records are easily available.
- ☐ Store the completed training documentation form according to regulatory requirements and company policy.
- ☐ Increasingly, training documentation is on-line. To guarantee proper documentation, all safety and health training records should be entered on-line.

Note the sample training documentation form on p. 94.

MEETS THE FOLLOWING MSSC PRODUCTION STANDARDS

P2-K3C	Training conducted is documented correctly and training records are updated and easily available.
P2-K3D	Training is relevant to equipment, tools, materials, and processes at the workstation.
P2-K3F	Training and training documentation meet all company and regulatory requirements.
P3-A1A	Input all safety and health training into data-base to guarantee proper documentation.
P3-A1C	Use computers to access training programs.
P3-A14E	Document safety incident and training orientation.
P6-K1E	Training outcomes are documented.
P6-O2I	Skill in keeping training records.
P6-A1C	Track training on database.
P8-O7A	Knowledge of documentation process to track and maintain training records and certifications.

(continued on next page)

Training Documentation Form

Personnel Information

Name:_____ **Signature:**_____

Department:_____

Position Title:_____ **Date:**_____

Date	Procedure	Person Trained		Trainer	
dd/mm	Equipment use/procedure	Name (BLOCK LETTERS)	Signature	Name (BLOCK LETTERS)	Signature

COPYRIGHT © GLENCOE/MCGRAW-HILL

Evaluating Job Training

A worker might be evaluated on the overall quality of the task that the worker performs. The training that worker receives might also be evaluated to ensure that the training is adequate to produce competent employees. Insufficient or inadequate training can lead to poor job performance.

Purpose of Training Evaluations

The purpose behind a training evaluation is to use the data collected in the process to improve the training materials and methods. By improving the training materials and methods, it is possible to ensure that employees are competent in performing their jobs. It will also improve the performance records of those who are already competent. At the very least, post-training evaluations should indicate that workers can operate equipment safely and that the training techniques are appropriate.

Training Evaluation Formats

Many different training evaluation formats are available. The format used should depend on the type and level of training that is to be evaluated as well as the data that will need to be collected. The evaluation formats can be tailored to different situations to maximize their effectiveness. Some of the common formats are:

- **Numerical rating scale.** It is used to evaluate a trainee on many tasks and helps control the subjectivity of the evaluator. It is often used to collect post-training feedback.
- **Questionnaire.** It is used to solicit feedback from trainees in the form of opinions.
- **Checklist.** It is used to measure job performance after training to evaluate whether or not the training increases job performance. It can also be used to determine if the training session was conducted properly.
- **Interview.** It is used when on-the-fly adjustments to the questions or in-depth review of the answers would be helpful.
- **Observation.** It is used when performance data is required long after the training has finished. Effectiveness depends on when the task is performed and the expertise of the observer.

Conducting Training Evaluations

Training evaluations should be conducted at a variety of different times and in a variety of different situations. This allows a more accurate picture of the effectiveness of the particular job training program. The type of training data that is required and the particular aspect of the training in question will help determine when the evaluation is conducted. The specific setting will depend on the level of the evaluation. Evaluations can be internal or external.

- Internal evaluations concentrate on reviewing course materials, trainee test data, trainee reaction to training, and instructor evaluations by training staff.
- External evaluations concentrate on reviewing the impact of the training on the job.

To conduct an effective job training evaluation, you must start by asking the right questions. In other words, you need to determine what questions the finished job evaluation should answer. The following questions are appropriate.

- What are the objectives of the job training program?
- Is the program accomplishing its objectives?
- What are the strengths of the training program?
- What are the weaknesses of the training program?
- Is the trainer skilled in delivering training?
- Is the training conducted in an effective and appropriate manner?
- Which trainees benefited the most and which the least?
- Is the program appropriate, given its intended purpose and target population?

(continued on next page)

Training needs should be assessed regularly. Listen to the concerns of staff. Workers may request training in weak areas to support growth and improve their work. Make sure that new requirements as well as current and future training issues are identified in a timely manner. Topics of training can be determined by reading maintenance manuals, participating in vendor training, and reviewing OSHA standards. The following checklist can be used to help maximize the use of the data collected in a training evaluation.

❑ Document quality and effectiveness appropriately.

❑ Organize training to meet the needs of the worker in order to maximize results.

❑ Give proper training to employees on new production processes.

❑ Ensure that training approaches effectively achieve training goals.

❑ Identify the need for appropriate cross-training, or training in different tasks or skills.

❑ Determine if training is necessary prior to starting the next project.

❑ Provide career development and training based on the data collected as well as team recommendations.

❑ Make suggestions regarding training materials and content to the appropriate parties.

❑ Utilize the knowledge gained to provide better training to other employees.

❑ Use the data to communicate safety, training, and job-specific needs to the appropriate parties.

Note the example of a trainee feedback evaluation form on p. 97.

(continued on next page)

COPYRIGHT © GLENCOE/McGRAW-HILL

Trainee Feedback Evaluation

COURSE/PROGRAM:_____   DATE:_____

NAME (Optional):_____   INSTRUCTOR'S NAME:_____

REVIEWED BY:_____   DATE:_____

Please rate the following statements using the following scale:

1   Strongly Disagree

2   Disagree

3   Neutral

4   Agree

5   Strongly Agree

1.	Time allotted to each unit of instruction was correct.	1 2 3 4 5
2.	Examples, analogies, and topics in training were relevant to your job needs.	1 2 3 4 5
3.	Training aids, audio-visuals, and handouts were current, accurate, and relevant to your job needs.	1 2 3 4 5
4.	As a result of attending the program or course, you are better prepared to perform your present duties.	1 2 3 4 5
5.	The classroom setting helped to promote learning.	1 2 3 4 5
6.	Facility specifics were taught where needed.	1 2 3 4 5
7.	The classroom training you received was beneficial to you in your understanding of facility operations.	1 2 3 4 5
8.	The information received in training was accurate and consistent with information received in the facility.	1 2 3 4 5
9.	The material was appropriate for your perspective (participant position, responsibilities, interests, beginning knowledge level).	1 2 3 4 5
10.	Your overall questions were answered satisfactorily.	1 2 3 4 5
11.	Overall, the course/program was beneficial and will help me perform my job.	1 2 3 4 5

ADDITIONAL COMMENTS:

(continued on next page)

COPYRIGHT © GLENCOE/McGRAW-HILL

MEETS THE FOLLOWING (MSSC) PRODUCTION STANDARDS

P1-A11B	Request training in weak areas to support growth and improve work.
P2-K3A	Training was conducted in an effective and appropriate manner.
P2-O5A	Skill in delivering training.
P5-K1	Communicate safety, training and job-specific needs.
P5-K1C	Current and future training issues are identified in a timely way.
P5-A5C	Organize training to meet the needs of the worker in order to maximize results.
P5-A12C	Present training sessions to employees on new production processes.
P5-A13B	Listen to concerns of staff to better provide training.
P6-K1	Provide training to other employees.
P6-K1A	Cross-training is provided as appropriate.
P6-K1B	Training needs are assessed regularly.
P6-K1C	New requirements and training issues are identified.
P6-K1D	Training approaches effectively achieve training goals.
P6-O2E	Skill in providing cross-training.
P6-A4D	Determine if training session is necessary prior to starting next project.
P6-A10B	Provide career development and training based on team recommendations.
P7-K1D	Suggestions regarding training materials and content are made to correct parties.
P7-K1E	Evaluations and feedback are utilized to improve training materials and methods.
P7-K1G	Post-training evaluation indicates that workers can operate equipment safely.
P7-K1H	Training and facilitation techniques used are appropriate for the audience.
P7-K1I	Quality and effectiveness of training are documented appropriately.
P7-A2A	Determine the topics of training (by reading maintenance manuals, participating in vendor training, and reviewing OSHA standards) to assess training needs.
P8-O7B	Knowledge of analytical methods for determining training needs (i.e., focus groups, structured interviews, surveys).

COPYRIGHT © GLENCOE/MCGRAW-HILL

Providing Excellent Customer Service

Successful businesses focus on developing customer loyalty among external customers. Loyal external customers may represent 80% of a company's business. Keeping such customers is often the key to the success of a business. The relationship of a business to its customers depends on achieving the following three goals.

- **Meet stated needs.** To meet customer needs, you must produce the product according to their specifications. This will also help prevent customer complaints.

- **Satisfy unstated needs.** To satisfy unstated customer needs, you must provide enhancements to the product or service. This will help develop consumer confidence.

- **Delight the customer.** To delight the customer, you need to provide outstanding service or pricing. This will help build customer loyalty.

To achieve these goals, you must have the right attitude, know your customer, improve customer service, and maintain customer contact.

Assess Your Attitude

The quality of customer service depends on your attitude and behavior. Answer the following questions to assess your attitude and behavior toward your customers.

- What impression do you make on a customer?
- Do customers refer your company to others?
- Do your customers return for repeat business?

Know Your Customer

- ❏ Know as much as you can about each customer's need for your products.
- ❏ Provide complete information about your current products.
- ❏ Present product information concisely and accurately in language the customer can easily understand.
- ❏ Read RFIs (requests for information) from the customer and paraphrase them into specific orders.

- ❏ Give your customer confidence that any help and support will be provided.
- ❏ Respond efficiently to issues and concerns raised by customers.
- ❏ Deliver what you promise. If you are unable to do what you promise in the required time, notify the customer and clearly explain why.

Improve Customer Service

To influence customers to purchase again from you, you need a product that satisfies customer needs at a fair price. You also need a good customer service policy. The following practices will help you improve your customer service.

- ❏ Make sure your customer service policy is centered on customer satisfaction.
- ❏ Actively encourage improvement in your company's customer service.
- ❏ Talk to your customers to identify their service concerns.
- ❏ Listen to what the customer wants. If you are unsure, ask questions.
- ❏ Make sure that your customer service is at the customer's convenience.
- ❏ Know how to respond to a customer's cultural needs and special needs.

Maintain Customer Contact

Maintaining customer contact will keep your products before the customer. Keeping in touch provides the following benefits to the manufacturer.

- Helps the manufacturer understand the customer's requirements for the product being made.
- Provides knowledge about the product's uses.
- Helps ensure that the customer's needs are met.

The following practices will help you maintain contact with your customers.

- ❏ Let the customer know you want to hear from them.

(continued on next page)

❑ Give the customer easy access to your company by providing comment cards and complete contact information, including e-mail addresses.

❑ Publish a business newsletter to promote new products and services and present testimonials from satisfied customers.

❑ Offer discounts for repeat business and incentives for bulk purchasing.

❑ Send reminder notices when it is time for your customers to reorder.

❑ Start a customer loyalty program to offer a reward for continued patronage.

MEETS THE FOLLOWING (MSSC) PRODUCTION STANDARDS

P4-A13D	Listen to customers to obtain knowledge of product uses.
P5-CWF	Communicate with co-workers and/or external customers to ensure production meets business requirements.
P5-K2	Communicate material specifications and delivery schedules.
P5-K2A	Communication reflects knowledge of material specifications.
P5-K2B	Delivery schedules are clearly communicated.
P5-K2C	Communication demonstrates knowledge of customer and business needs.
P5-K4C	All parties are notified of production issues and problems in a timely way.
P5-K4D	Communication demonstrates knowledge of customer and business needs.
P5-O4C	Knowledge of customer and business needs in order to communicate effectively.
P5-A5D	Organize and plan routine communication with customers.
P5-A15C	Read RFIs and their related materials and paraphrase into specific orders/requests of co-workers and clients.

COPYRIGHT © GLENCOE/McGRAW-HILL

Reading Views on Technical Drawings

Technical drawings are graphic representations of a product that include dimensions and all other information necessary to manufacture the product. Because technical drawings are two-dimensional, several drawings may be required to describe a product completely.

Linetypes

To understand a technical drawing, you must first understand the meaning of the different types of lines. **Fig. 1** shows the four basic types of lines. Continuous lines are used for object outlines and features such as holes and slots. Hidden lines show features that are hidden in a particular view of an object. Centerlines are used to locate the centers of holes and other round features and to show symmetry. Phantom lines show alternate positions of parts, among other things.

■ **Fig. 1**

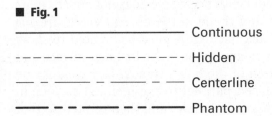

Multiview Drawings

Many technical drawings show multiple views of a part or product in order to describe it completely. These drawings are called *multiview drawings*. The basic views in a multiview drawing are the front, top, right-side, left-side, back, and bottom views. In the front view, you see the object as if you were looking directly at the front of it. In the left-side view, you see the object as if you were looking directly at its left side, and so on. Only the views necessary to describe the object completely are included in the drawing.

The number of views required to describe a product completely is determined by the complexity of the product. A simple product such as the head gasket shown in **Fig. 2** needs only one view with a note to define the thickness. A bracket like the one shown in **Fig. 3** requires a front view and

a top view. The slightly more complex bracket shown in **Fig. 4** requires three views: front, top, and right-side.

■ **Fig. 2**

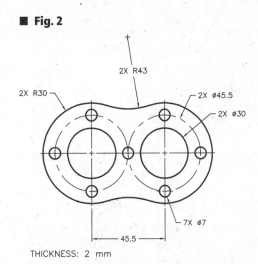

THICKNESS: 2 mm

■ **Fig. 3**

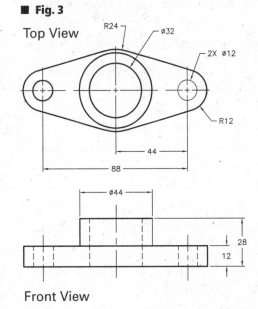

Top View

Front View

(continued on next page)

■ Fig. 4

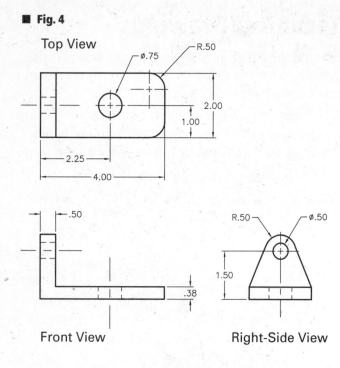

Top View

Front View   Right-Side View

Interpreting the Normal Views

To read a multiview drawing, you must know the conventions for placing the views. Notice the arrangement of the views in the three-view drawing in **Fig. 4**. The top view is always placed above the front view and is aligned with it. The right-side view is always placed to the right of the front view and is also aligned with the front view. These three views are known as the *normal views* because they are so commonly used.

As you look at multiview drawings, try to think of how the finished product will look. This practice, called *visualization*, is an important skill for reading technical drawings. Visualization requires practice; the more you try to visualize objects based on their multiview drawings, the better you will be able to "see" them.

Interpreting Auxiliary Views

Some products require more than just the three normal views. For example, the bracket in **Fig. 5** contains an inclined surface with a hole. The location of the hole cannot be dimensioned in any of the normal views because it does not appear in its true size and shape in the normal views. In this case, an auxiliary view must be created to show the true size and shape of the inclined surface and the location of the hole. An auxiliary view is one that is placed parallel to the inclined surface in one of the normal views and then rotated so that the surface appears at its true shape and size.

■ Fig. 5

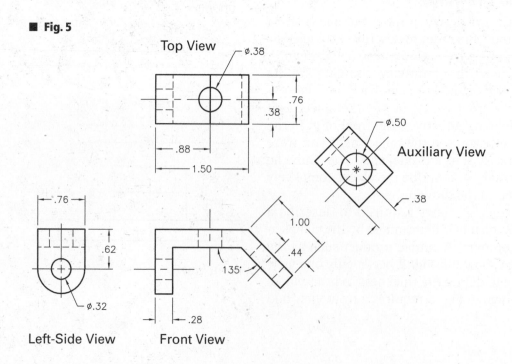

Top View

Auxiliary View

Left-Side View   Front View

(continued on next page)

COPYRIGHT © GLENCOE/MCGRAW-HILL

■ **Fig. 6**

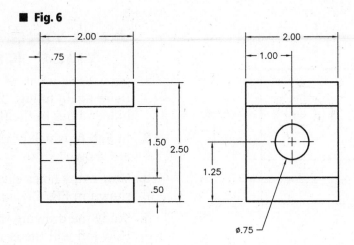

■ **Fig. 7**

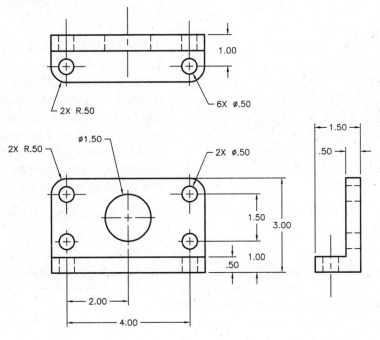

(continued on next page)

COPYRIGHT © GLENCOE/MCGRAW-HILL

■ **Fig. 8**

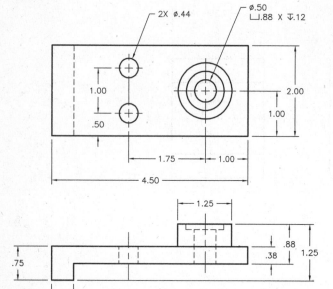

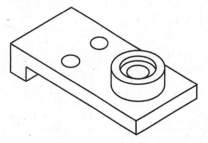

■ **Fig. 9**

APPLYING YOUR KNOWLEDGE

1. Refer again to **Fig. 3**. What do the hidden lines in the front view indicate?
2. In **Fig. 6**, which of the normal views are shown?
3. How many holes exist in the drawing shown in **Fig. 7**?
4. Study the drawing shown in **Fig. 8**. How tall will the finished product be? How wide will it be? How deep will it be?
5. How many views will it take to describe the part shown in **Fig. 9** completely? Which views would you choose?

PRODUCTION CHALLENGE

Your work at a fine wood furniture manufacturer involves using technical drawings to create a newly designed entertainment unit. The unit includes a number of compartments and shelves as well as two sliding doors and two hinged doors. Which type(s) of views would you need to see to make sure that the end product follows the design?

MEETS THE FOLLOWING (MSSC) PRODUCTION STANDARD

| P1-06D | Knowledge of how to use diagrams and technical drawings. |

COPYRIGHT © GLENCOE/MCGRAW-HILL

Interpreting Drawings to Meet Customer Needs

Technical drawings include all of the information necessary to manufacture a product. To meet customer needs effectively, team members must therefore have a basic understanding of the symbols and other conventions used on technical drawings. The team can then gain consensus about producing the product and avoid future nonconformances.

Symbols

To keep complex drawings from becoming cluttered and confusing, standard symbols are often used instead of writing out specifications. Some of the symbols that are commonly used on technical drawings are shown in **Fig. 1**. Failure to notice or understand these symbols can result in a product being manufactured or finished incorrectly.

■ **Fig. 1**

Symbol	Meaning
℄	Centerline: Identifies the center of a circle or other feature; also used to show symmetry
⊔	Counterbore: Specifies a cylindrical bore of specified depth and diameter to accommodate the head of a screw or other fastener
∨	Countersink: Specifies a conical hole of specified depth and diameter to accommodate the head of a screw or other fastener
⊤	Depth: Indicates that the following number is the depth of a blind hole (one that does not extend all the way through a part), counterbore, or countersink
✓	Finish mark: A general-purpose finish symbol that indicates that a surface is to be machined or finished
R	Radius: Indicates that the following number is a radial dimension (half of the diameter)
⌀	Diameter: Indicates that the following number is a diameter dimension

Tolerances

No machine tool in existence today is absolutely accurate. Therefore, instead of specifying an absolute dimension, designers build a specific tolerance into each dimension on a technical drawing. The tolerance specifies the upper and lower limits within which the part or feature must be created. For some parts or features of parts, the tolerance can be loose, allowing a large degree of error. For others, however, a much tighter tolerance is required. In general, tolerances are set to satisfy the requirements of the customer.

Tolerance vs. Cost

It is important to note that as tolerances become tighter, cost rises rapidly. A common rule of thumb is that every time you divide a tolerance in two, you increase the price of the operation by four times. For example, suppose it costs $1.25 to create a hole at a tolerance of ±.010. Changing the tolerance to ±.005 increases the cost of this operation to $5.00. Therefore, tolerances are usually set as loose as possible while still meeting the customer's requirements.

Types of Tolerances

Every dimension on a technical drawing must be toleranced. Individual tolerances can be shown on a drawing in several different styles, depending on the tolerance requirements and the preferences of the drafter. Some of the most common styles are shown in **Fig. 2**. However, it would be awkward and inconvenient to specify a tolerance for each dimension individually, so drafters either use a general drawing note or a special tolerance box to specify a general tolerance. This tolerance applies to all dimensions not otherwise toleranced on the drawing. In **Fig. 3**, the drafter has placed a note in the lower left corner of the drawing.

■ **Fig. 2**

Tolerance Style	Meaning
12.250±.005	Indicates a bilateral tolerance (one in which the upper tolerance is the same as the lower tolerance)
12.250 +.005 −.002	Indicates a tolerance in which the upper and lower values are different
12.252 12.248	Specifies only the largest and smallest sizes allowed; the tolerance is the difference between the limits

(continued on next page)

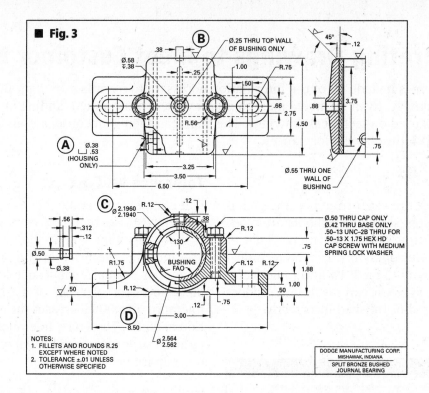

■ **Fig. 3**

APPLYING YOUR KNOWLEDGE

Refer to the dimensions at the circled letters in **Fig. 3** to answer these questions.

1. What is the meaning of dimension **A**?

2. What is the meaning of symbol **B**?

3. A checker in Quality Control measures the hole dimensioned at **C** and finds that the actual diameter of the hole is 2.1945. Will this diameter meet the customer's specifications?

4. The checker measures the dimension at **D** and discovers the actual width to be 8.48. Will this width meet the customer's specifications?

PRODUCTION CHALLENGE

Five of your customer's specifications for an upholstered armchair are unclear. There is a severe shortage of one of the specified hardwoods. One of the operations requires a machine that is being repaired. How would you resolve these problems?

MEETS THE FOLLOWING (MSSC) PRODUCTION STANDARDS

P1-03A	Skill in interpreting technical drawings so that customer needs are met.
P1-A15B	Read blueprints to meet customer needs.
P8-05B	Skill in analyzing technical data and drawings and gaining group consensus to avoid future nonconformances.

COPYRIGHT © GLENCOE/MCGRAW-HILL

Obtaining Customer Feedback

Feedback is a form of communication that customers can use to inform manufacturers of how effectively they are meeting customer needs. When manufacturers learn of a problem a customer has with their products, they focus on resolving it. The more prepared your organization is to receive customer feedback, the more successful it will be in acting on it. Everyone in your organization should be prepared to listen to customer requirements, complaints, and praise. They should be willing to forward this information to the appropriate people. Information is usually given to production so that all team members can work on resolving the problem.

Companies might have different policies and procedures for handling customer complaints or finding out how well their products are performing. However, there are three ways that most companies approach when obtaining customer feedback.

- Encourage customers to provide feedback.
- Plan for customer feedback so you know what to expect.
- If negative feedback is obtained, follow through and resolve the problem.

Encouraging Customer Feedback

Companies should not wait to hear from customers. They should always keep the lines of communication open by maintaining effective customer contact. Following these basic guidelines will encourage customer feedback.

- ❏ Encourage feedback by promoting a company policy of caring for and delighting the customer.
- ❏ Have procedures in place to gather feedback from customers about how your company addresses their needs.
- ❏ Encourage employees to obtain customer feedback by observation, active listening, and problem solving.
- ❏ Gauge the attitudes for receiving and using customer feedback by asking the following questions.
  - Do managers and employees intend to pay attention to customer feedback and act on it?
  - Are all managers and employees committed to taking action based on customers' input?
  - Do all managers and employees accept that the successful use of customer feedback might result in change and added work?

Planning for Customer Feedback

Customer feedback can be planned or unplanned. Unplanned customer feedback is received in the form of product complaints. It also is received in the form of a recommendation of your product by a satisfied user.

Planned customer feedback is usually obtained through highly organized customer surveys. Print versions of customer surveys are usually in the form of questionnaires or comment cards. Surveys can also be conducted by telephone, e-mail, or computer.

A campaign to obtain customer feedback in this way must be carefully planned. The following guidelines outline the basic considerations.

- ❏ Determine the frequency of your campaigns to obtain customer feedback.
  - Is your need for feedback seasonal?
  - Has a critical event made feedback more important?
- ❏ Determine your use of customer feedback. Will it be used to:
  - Assess past business performance?
  - Forecast future sales?
  - Revise, correct, or improve a process?
  - Identify customer needs and expectations?
  - Identify customer needs that are not being met so that such needs can be addressed proactively?
  - Assess customer quality expectations and other key concerns?
  - Review customer needs on a regular basis?
  - Improve customer relationships?
  - Obtain information about the product's use?
  - Make decisions regarding product design?

(continued on next page)

- Assess field failure of products?
- Gain information regarding production and scheduling?
- Assess consumer trends as they relate to your product?

Above all, it is crucial to ask specific questions about the customer's experiences with the company. These questions should address both the product and customer service. By taking responses seriously, the company can improve their products and procedures. Questions can include the following.

- Did the product meet your specifications?
- Did the product meet your quality expectations?
- Was the product delivered on time?
- Was the product packaged and shipped appropriately?
- Were issues preventing your needs from being met addressed proactively?
- What can we do to meet your needs more consistently?

Resolving Customer Complaints

Customer complaints are negative feedback. Respond to customer complaints in a manner that reflects favorably on your company. Follow these guidelines.

- ❑ Make it easy for customers to file a complaint.
- ❑ Listen carefully to the customer filing the complaint.
- ❑ Ask questions to clarify the complaint to ensure that you understand it.
- ❑ Let the customer know you understand the complaint.
- ❑ Do not apportion blame.
- ❑ Thank the customer for advising you of the problem.
- ❑ Tell the customer what action you will take to resolve the complaint.
- ❑ Identify a time by which the complaint will be resolved.
- ❑ Deal with complaints promptly and efficiently.
- ❑ Ensure that all promised action is completed promptly.
- ❑ Notify the customer of the action taken.
- ❑ Document the complaint and the action taken to resolve it.
- ❑ Analyze the information.
- ❑ Pass on the information to the appropriate individuals, who can assess it for use in improving the product or procedures.

MEETS THE FOLLOWING (MSSC) PRODUCTION STANDARDS

P1-K1	Identify customer needs.
P1-K1C	Customer needs are reviewed on a regular basis.
P1-K1F	Issues preventing customer needs from being met are addressed proactively.
P1-A13B	Listen to customer requirements, complaints and praise and forward the information to appropriate people.
P4-A13D	Listen to customers to obtain knowledge of product uses.
P5-K2C	Communication demonstrates knowledge of customer and business needs.
P5-A11C	Share knowledge and ask for feedback from customers.
P8-O9A	Knowledge of customer quality expectations and other key concerns.
P8-A13A	Receive feedback from customer regarding field failures in order to promote future business.

Copyright © Glencoe/McGraw-Hill

Scheduling Meetings

The skills required to schedule a face-to-face meeting differ from those required to conduct a meeting. Meetings vary in complexity. Several factors must be considered. Following are basic guidelines for scheduling a face-to-face meeting and following up.

Identifying the Attendees, Time & Place

1. Determine whether those invited will be mainly external customers, internal customers, or both. This may determine the complexity involved in scheduling the meeting.

2. Contact the customer to explain the benefits of holding the meeting. Ask if they would like to attend the meeting.

3. Ask the customer to identify the topics they would like to discuss at the meeting.

4. Identify the topics you would like to discuss. These should include customer needs, requirements, and expectations. Topics should also include a discussion of specifications, production schedules, and timelines.

5. Ask the customer to identify those they would need to invite to the meeting. Mention that only those involved in the project should be invited. State that meetings are often more productive if the number invited is limited.

6. Inform the customer of those you plan to invite.

7. Obtain contact information for those to be invited to the meeting.

8. Set a tentative time and place for the meeting after consulting with the customer.

9. Set a time for beginning the meeting and a time for ending the meeting.

Reserving the Space

10. Reserve the meeting room. Will the meeting be held on company premises or locally at another location? Will the meeting be held out of town? In either case, meeting rooms will need to be reserved.

11. Make arrangements for the needed number of tables and chairs to be placed in the meeting room. Specify the table arrangements and the number of chairs per table. Will a lectern be needed?

12. Arrange for visual aids if needed. If projectors and other nonbattery-powered electrical devices are needed, check that electrical outlets are close to where the devices will be placed.

13. Arrange for food and beverages if needed. Specify where and at what time the food and beverages should be served. Check to see if any of those attending have special dietary needs.

Sending Invitations & Agendas

14. Prepare an agenda for the meeting. An agenda is a printed list of topics that you and the customer wish to discuss at the meeting. List the topics in order of importance.

15. Send all interested parties an invitation to the meeting and an agenda as well as any other information needed to prepare for the meeting. This can usually be done by e-mail.

16. Ask those invited to accept or decline the invitation. This also can be done by e-mail. If they decline, ask them to suggest dates and time when they would be available.

17. If necessary, make travel arrangements, including hotel reservations, for those attending the meeting.

(continued on next page)

Rechecking

18. Check that all key people can attend.

19. Confirm meeting room reservations, travel reservations, and hotel reservations.

20. Consider rescheduling the meeting at an alternative time and place if key people cannot attend. You might also consider teleconferencing.

Following Up

21. Organize and plan routine communication with those invited to the meeting.

22. If travel expenses are to be reimbursed, ensure that this is done.

23. Send information presented at the meeting to those scheduled attendees who were unable to attend.

Considering Alternatives to Face-to-Face Meetings

- Telephone conference calls have become substitutes for formal meetings. This is known as teleconferencing.

- In video teleconferencing, a video screen projects the images of those you are talking to. Your image is also shown to them.

MEETS THE FOLLOWING PRODUCTION STANDARD

P1-A6A	Meet with customers in a professional way to identify customer specific requirements.

COPYRIGHT © GLENCOE/MCGRAW-HILL

Conducting Meetings

Meetings are important in manufacturing. Your team may need to work closely with the heads of other departments to plan and implement company goals. For example, your team may need to meet with the production manager and the procurement department to ensure that plant inventories are adequate for the job at hand. This is vital to the successful completion of a production job.

A meeting signals to the customer—whether external or internal—your team's interest in their concerns. It provides an opportunity to clarify information and ensure that all concerned have the information they need. It provides an excellent opportunity for providing project status updates. Successful production depends on effective scheduling, staffing, procurement and maintenance of equipment, quality control, inventory control, and the coordination of production activities. It allows you to resolve issues regarding production and process. You might need to meet with:

- External customers regarding their order.
- Contractors to discuss material delivery challenges.
- Production team to discuss daily production issues.

Following are basic guidelines for conducting a meeting in a professional manner.

Focusing & Facilitating

1. Make sure any needed visual aids are in place.

2. Make sure you have the information you need to present. Such communication will demonstrate your knowledge of the customer and business needs. For example, you may have obtained information about customer needs from their e-mails. You may then have used this information as a basis for production scheduling.

3. Make sure you have enough copies of any handouts.

4. Open the meeting by introducing yourself.

5. Ask the participants to introduce themselves.

6. Check that everyone has a copy of the agenda and any needed handouts.

7. Identify problems that need to be solved, goals that need to be set, and plans that need to be made.

8. Guide the discussion so that it remains focused on the topics stated in the agenda.

9. Identify matters that require a specific action to bring about a solution.

10. Listen objectively to the company's specific requirements, even though they may exceed existing specs.

11. Discuss product aspects and printed specifications to ensure understanding of needs. Pay special attention to:
    - Customer requirements.
    - Customer needs.
    - Customer quality expectations.
    - Customer use of the product. The answer will help confirm that the product as built will meet customer needs.
    - Any other key customer concerns.

12. Discuss customer specifications.
    - Are the customer specifications up-to-date?
    - Do the customer requirements exceed the specs?
    - Are any deviations from the customer specifications acceptable?
    - If there are deviations, have the deviations been identified?

13. Present a tentative production schedule based on priorities and available resources.

14. Ask the customer if the production schedule is acceptable. If it is not, set a deadline for establishing an acceptable production schedule. This may require a change in priorities or a shift in resources.

(continued on next page)

15. Throughout, be sure to demonstrate sensitivity to the customer's changing delivery schedule.

16. Present a tentative timeline based on priorities and available resources.

17. Ask the customer if the timeline is acceptable. If it is not, agree on a deadline for establishing an acceptable timeline.

18. Take careful notes. You will need to share the information with others.

19. At the end of the meeting, tell the customer that you will keep them informed of the status of the project. Tell them how frequently they may expect this information and how it will be delivered.

Following Up

20. As necessary, give accountable individuals the authority to follow up.

21. Realize that internal and external customers share certain needs. For example, every customer wants quality work done on time. Each customer, however, has specific needs. In following up, you need to be attentive to these specific needs.

22. Organize and plan routine communication with the customer. You will probably want to give them regular progress updates on production and shipment.

23. Inform your peers of the reviewed customer product specifications and requirements. Be sure to explain the need for any added production processes.

24. Inform needed parties of revised production and shipping deadlines. Be sure to explain anything that might be misunderstood.

25. Make sure that all customer needs are communicated effectively to others. This includes communication between one shift and another, as well as communication with managers and co-workers.

26. Verbally clarify customer needs to co-workers. Make sure that they understand.

27. Record meeting notes.

28. Distribute meeting notes to interested parties for review and update.

29. Ensure that documentation relating to customers is distributed to the appropriate parties in a timely manner.

MEETS THE FOLLOWING (MSSC) PRODUCTION STANDARDS

P1-K1A	The different and common needs of internal and external customers are recognized.
P1-K1B	Customer contact about product aspects and printed specifications is maintained to ensure understanding of needs.
P1-K1D	Customer specifications are up-to-date.
P1-K1E	Customer needs are communicated effectively to others including shift-to-shift, co-workers, and managers.
P1-A6A	Meet with customers in a professional way to identify customer specific requirements.
P1-A9B	Lead team meetings to determine customer needs, set up priorities and available resources.
P1-A10D	Resolve issues regarding production and process before implementation.
P1-A12D	Discuss schedules and establish timelines with customers.
P1-A12E	Discuss needed changes in materials with customers.
P1-A13D	Listen objectively to the customer's requirements, even though they may exceed existing specs.
P4-A6B	Meet with fellow employees and discuss business needs in a professional manner.
P5-K4D	Communication demonstrates knowledge of customer and business needs.
P5-A6C	Meet with contractors in a professional manner to discuss challenges.
P5-A7D	Demonstrate sensitivity to customer's changing delivery schedule.
P6-A9A	Lead daily production meetings.
P6-A14A	Record meeting/team notes and publish polished copy for review and update.
P8-K3D	Documentation required for customers is distributed to appropriate parties.

   COPYRIGHT © GLENCOE/MCGRAW-HILL

Understanding Geometric Shapes

All discrete products, from the simplest to the most complex, can be broken down into basic geometric shapes. Understanding these shapes can help you analyze both the product and the blueprint for that product. Some of the geometric shapes commonly found in products are shown in **Fig. 1**.

■ Fig. 1

Planar Shapes	Three-Dimensional (3D) Shapes		
Triangle	Square Pyramid	Triangular Prism	Triangular Pyramid
Square	Cube	Hexahedron (Box)	
Pentagon	Pentagonal Prism		
Hexagon	Hexagonal Prism		
Circle	Cylinder	Cone	Sphere
Ellipse	Elliptical Cylinder		

Analyzing a Part

Before production can begin, a process sheet is often generated to list, in order, the operations required to create a part or product. To know which operations are needed, the designer or engineer looks at the part as a sum of its basic geometric shapes. By analyzing the bracket geometrically, the designer or engineer can plan the drilling, forming, welding, or casting steps that will be needed to manufacture it.

For example, the pole bracket shown on the right in **Fig. 2** can be broken into the basic geometric shapes shown on the left. Notice that the holes can be thought of as cylinders that need to be subtracted from the part. Cylinders that are removed from a part to create holes are sometimes referred to as *negative cylinders*.

■ Fig. 2

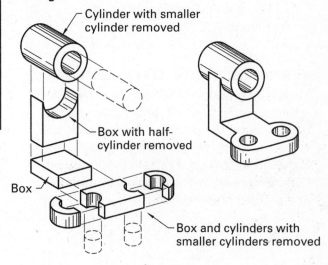

Cylinder with smaller cylinder removed

Box with half-cylinder removed

Box

Box and cylinders with smaller cylinders removed

(continued on next page)

COPYRIGHT © GLENCOE/McGRAW-HILL

Interpreting an Exploded Assembly Drawing

If a product is made up of several different parts, the set of blueprints developed to describe the product often contains an exploded assembly drawing similar to the one shown in **Fig. 3**. This is a pictorial drawing that shows every part in its proper place in the assembly but "exploded" apart to clarify how the parts fit together.

An exploded assembly drawing can help you understand the geometry underlying a product. For example, in the depth gauge shown in **Fig. 3**, several cylinders are evident. The base can be thought of as a box with two triangular prisms removed from it.

APPLYING YOUR KNOWLEDGE

1. Could the base of the depth gauge shown in **Fig. 3** be created from a square pyramid shape?

2. What kind of operation might be used to create the holes in the pole bracket shown in **Fig. 2**?

3. How might the base for the depth gauge in **Fig. 3** be manufactured? List the operations.

PRODUCTION CHALLENGE

Most food product containers are shaped like squares, rectangles, or cylinders. Several factors influence the manufacturer's choice of a container. Identify reasons that a certain shape is chosen to contain a certain food product. Consider all aspects of packaging, distribution, and product presentation.

■ **Fig. 3**

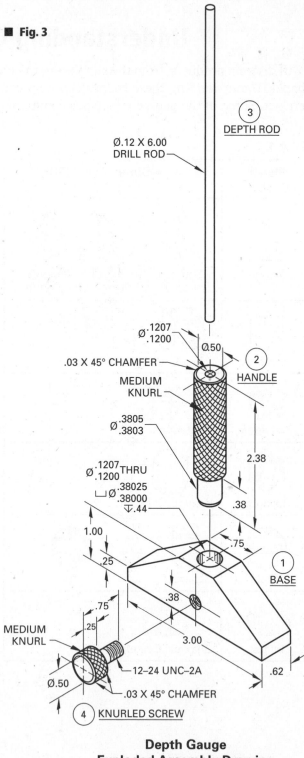

**Depth Gauge
Exploded Assembly Drawing**

MEETS THE FOLLOWING (MSSC) PRODUCTION STANDARD

P1-A16D   Understand geometry in order to interpret blueprints.

COPYRIGHT © GLENCOE/McGRAW-HILL

Using CAD Design & Production Techniques

Traditionally, companies worked with sets of two-dimensional (2D) blueprints or technical drawings to design a product, manufacture it, and check the quality of the manufactured items. However, designing in 2D was a long and painstaking process. Even the smallest change in design had to be carried out in every drawing in the set. Changing the drawings required much skill and experience. Larger changes sometimes required recreating some of the drawings from scratch. In addition, designers were often frustrated because no sooner had they made one change to the set of drawings than another change would be required.

Designing in CAD

The development of computer-aided drafting (CAD) has changed all that. Even the development of 2D drawings is more efficient than the traditional drafting methods because changes can be made quickly and easily in the electronic CAD file. This speeds up the design cycle, allowing manufacturers to produce parts in a more timely manner.

Today, however, designers and engineers often work directly on three-dimensional (3D) CAD models. The 2D drawings required for manufacturing can be extracted directly from the models. In some cases, the models are fed into conversion programs that create the code necessary to drive computer-aided manufacturing (CAM) equipment. When this method is used, the traditional 2D drawings are sometimes not even required.

Rapid Prototyping

Another advantage of designing in CAD is that a model can be sent directly to stereolithography and other rapid prototyping machines. The result is a physical prototype of the actual part. This process is much faster and less labor-intensive than building a prototype manually. The chess pieces in **Fig. 1** are examples of the intricate detail possible using rapid prototyping techniques.

Rapid prototypes are commonly used by designers and engineers for several purposes, including:

- Showing a customer what the finished part will look like.

- Obtaining management approval of a design.

- Fitting with adjacent parts to verify fits and tolerances.

- Instant manufacturing to fulfill a customer order for a one-of-a-kind or rarely needed part.

■ **Fig. 1**
Courtesy of Arnold & Brown

Product Analysis

Solid models developed in CAD can also be used directly for testing and analysis. Instead of having to design or rent a wind tunnel, for example, the designer or engineer can subject the model to electronic analysis. The results are fast, reliable, and after the initial cost of the analysis software has been recovered, much less expensive than traditional analysis. Many different kinds of tests can be performed, including complex analyses such as finite element analysis (FEA) and computational fluid dynamics (CFD).

All of this is possible because CAD programs allow the designer to assign various materials, including many different woods, metals, glass, and plastics, to solid models. Objects to which these materials have been assigned take on the material conditions and mass properties of the individual materials. By testing models to which the intended materials have been applied, designers can discover weaknesses or design flaws that otherwise may not have been noticed until much further along in the product development.

(continued on next page)

Checking Clearances in Assemblies

In products that contain more than one part that must fit together, CAD models are often used to check for both fit and interference. No prototype is needed because the models exist in 3D space within the CAD program. The assembly in **Fig. 2** shows part of a CAD file that contains several models. Each part is an individual piece that can be moved independently of the others. By moving the parts into position and using the appropriate CAD commands, the designer can check fits, clearances, and interferences.

■ **Fig. 2**

Model by Timothy M. Looney

CAD/CAM

When CAD models are used in CAD/CAM settings to drive computer numerical control (CNC) equipment, even the processes used to create the part can be optimized. The computer determines the most efficient and economical tool paths to produce the part represented by the model. The computer also increases the consistency of the manufacturing process. This, in turn, increases the quality of the product.

APPLYING YOUR KNOWLEDGE

1. A design team is working on a new housing design for one of the company's most popular motors. The president of the company wants to see the team's ideas. The design team is working in CAD. What options do team members have for showing the president their designs?

2. The owner of a small manufacturing company that does not currently use CAD/CAM has asked for a report on the advantages and disadvantages of converting to a CAD/CAM system. What points would you make in such a report?

3. A company that manufactures kitchen and bath fixtures has decided to produce a new, high-end product line. The company currently uses CAD/CAM procedures and equipment. How might the company proceed to design the new product line?

PRODUCTION CHALLENGE

You are working for a company that produces diodes for optical fibers. How could you use CAD to explain to a new employee the difference between single-mode and multimode optical fibers?

MEETS THE FOLLOWING PRODUCTION STANDARD

| **P4-A1B** | Use CAD to design better fixtures and processes to produce quality parts in a timely fashion. |

 COPYRIGHT © GLENCOE/MCGRAW-HILL

Brainstorming

Some problems have simple solutions. If you want to sit at a table but all of the chairs are in another room, the obvious solution is to get a chair from that room. However, a solution to a problem is not always obvious. In such cases, you need a method for developing solutions. The method should be effective in producing solutions in the widest variety of cases. One such method is brainstorming.

Brainstorming is a method of problem solving that is used to generate ideas. You would then build consensus on the ideas presented by discussing them with the group.

Brainstorming involves two or more people working with each other to present as many ideas as possible. A group can produce more ideas than any one person working alone. The focus of brainstorming is on the *quantity* of ideas presented rather than on the quality of those ideas. By presenting as many ideas as possible, you increase the chances of generating a creative and useful solution. Each new idea that you present could also trigger an idea from another person in the group.

Rules for Brainstorming

Certain rules must be followed when brainstorming. These rules foster the environment needed for successful brainstorming. The creative process of brainstorming might fail if the following rules are not observed.

❏ Do not hold back on expressing ideas, even though they may seem silly or even absurd. Practice freewheeling, which is the spontaneous, free expression of ideas in a brainstorming session.

❏ Get to the point. State your idea quickly. Then yield the floor to another.

❏ Do not discuss ideas during brainstorming. Discuss the ideas afterwards.

❏ Do not criticize ideas.

❏ Build upon the ideas of others.

❏ Write down all ideas where they can be easily read by all participants.

A brainstorming session is no time to be negative. There are certain phrases that no one should say:

- It won't work.
- It's too expensive.
- We've never done that before.
- We've already tried that.

Brainstorming Exercise

Brainstorming is a relatively simple process. It involves two or more people generating ideas or solutions that address a common question or problem. As long as the rules are followed, brainstorming can be a productive experience for all involved. The following exercise will give you firsthand experience with brainstorming.

1. Gather a small group of people, preferably 3 to 6 individuals.

2. Designate a recorder. This person will be in charge of writing everyone's ideas in a prominent location.

3. Make sure that everyone can see the recorder's notes.

4. Identify the problem that needs to be solved. You might brainstorm a list of possible improvements to make as part of the continuous improvement process. You might also brainstorm to build consensus on the importance of inspection to determine quality and/or condition.

5. Organize the session so that everyone has the opportunity to give input. Take turns or provide alternate means of response. Encourage everyone to respond as much as possible.

(continued on next page)

6. When an idea is generated, the recorder should write it down.

7. Eliminate or combine duplicate ideas.

8. Discuss the ideas and determine which ones are worth further investigation.

Brainstorming can be used for a variety of tasks. Some of these include:

- Generating, prioritizing, and building consensus on continuous improvement projects.
- Generating ideas for an assigned project.
- Creating a list of design features and components.

- Creative problem solving.
- Identifying strengths and weaknesses.
- Generating solutions to team problems or project problems.

For extra practice, follow the steps above and have teams write procedures on the safe use of equipment in the workplace. Are there any existing safety problems? Can the work environment be made safer by making simple changes? Make sure that everyone gives input and that no one's input is criticized during the brainstorming process.

MEETS THE FOLLOWING PRODUCTION STANDARDS

P4-A10A	Build consensus by brainstorming all potential C.I. Projects in order to prioritize them and begin implementation.
P4-A10D	Build consensus on the importance of inspection to determine quality or condition.
P7-A8B	Have teams write procedures on safe use of equipment in the workplace.

COPYRIGHT © GLENCOE/MCGRAW-HILL

Setting Up for a Test Run

Acquiring training in setup and maintenance can improve your value as an employee. The following operations are the general steps for setting up a CNC machine tool and test-running a program. Refer to your CNC machine tool operation manual for specific information.

1. Before every step, ensure that you are wearing the appropriate personal protective equipment and ensure that all safety equipment is in place.

2. Identify the key information on the work order. A work order is a document authorizing the completion of a specific task. Not all of the information on the following checklist may be present on the work order.
   - ❑ Work order number.
   - ❑ Part name.
   - ❑ Material specifications.
   - ❑ Required operations.
   - ❑ Required number of parts.
   - ❑ CNC program number.
   - ❑ Production schedule.
   - ❑ Time allotted for setup.
   - ❑ Time allotted for production operations.
   - ❑ Prior operations.
   - ❑ Subsequent operations.
   - ❑ Quality control information.
   - ❑ Attached blueprint of part.

3. Prepare for machine setup.
   - ❑ Clean the machine and the work area.
   - ❑ Check all fluid levels. Adjust if necessary.
   - ❑ Familiarize yourself with the blueprint.
   - ❑ Obtain tooling, applicable gauges, and material.
   - ❑ Install the tooling and fixturing according to instructions on setup sheet.

4. Read machine setup sheet.
   - ❑ Make sure that the setup meets equipment specifications.
   - ❑ Make sure that the setup meets process specifications for internal and external customers.
   - ❑ Make sure that the manufacturing process cycle time meets the customer and business needs.
   - ❑ Identify the part with the operation.
   - ❑ Identify the cutting tools required.
   - ❑ Identify the tooling already set up and the tooling that needs to be set up.
   - ❑ Match the tool position to the program code.
   - ❑ Study the sequence of operations.
   - ❑ Review instructions.
   - ❑ Identify the location of program zero in relationship to workplace and fixture.

5. Prepare the workstation for production.
   - ❑ Clean the work area.
   - ❑ Clean the machine tool.
   - ❑ Check any coolant filtering devices.
   - ❑ Turn on machine tool.
   - ❑ Initialize the axes and control unit.
   - ❑ Check all fluid levels. Adjust if necessary.
   - ❑ Shut down machine tool.

(continued on next page)

6. Set up the tool.
   ❑ Review the operation manual for instructions regarding:
      - Defining the part zero position.
      - Tooling setup.
      - Offsets for specific machine tools.
      - How to turn off the machine in case of emergency.
   ❑ Turn on the machine.
   ❑ Initialize the axes and control unit.
   ❑ Mount the appropriate fixture or work-holding device to the machine tool table.
   ❑ Mount an edge-finding device in the machine spindle.
   ❑ Identify the zero points in the X-axis and Y-axis to the control. Follow the procedures in the CNC machine tool operation manual.
   ❑ Relate the axis positions to a fixture offset. Refer to the CNC machine tool operation manual.
   ❑ Secure the cutting tool in the appropriate tool holder.
   ❑ Mount the tool holder in the machine tool.
   ❑ Mount the workpiece in the fixture or work-holding device.
   ❑ Move the selected tool close to a point above the workpiece, using the manual-jog mode.
   ❑ Manually move the Z-axis to a point of contact with the top of the workpiece. Follow the procedures outlined in the CNC operation manual.
   ❑ Verify that the Z-axis offset for the specified tool is set to zero in the control unit.
   ❑ Ensure that no other offsets or data are set in the control unit for the tools being used in the operation.
   ❑ Document the setup procedures for repeatability according to company policy.

7. Load the program.
   ❑ Review the operation manual regarding program transfer.
   ❑ Transfer the program to the CNC machine tool.
   ❑ Activate the program.

8. Test-run the program.
   ❑ The program can be test-run with or without the workpiece.

Test-Run Without the Workpiece

❑ Make certain that no components are in the tool path.
❑ Review the machine operation manual regarding the test run.
❑ Select dry-run mode.
❑ Reduce feed rate override to zero.
❑ Reset control, which sets control and program to the beginning.
❑ Select cycle start.
❑ Step through each sequence of the program. Check that:
   - Tool path is correct.
   - There is sufficient clearance for tool approach.
   - There is sufficient clearance for movement through the workpiece.
   - There is sufficient clearance for tool exit and return to home position.
❑ If errors are found in the program code:
   - Stop the program.
   - Edit the program to correct the errors.
   - Restart the program.
   - Verify the edits.
   - Continue to prove the program.

(continued on next page)

   COPYRIGHT © GLENCOE/McGRAW-HILL

Test-Run with the Workpiece

❑ Load the program.

❑ Install the tooling and offsets assigned in the control.

❑ Install fixtures.

❑ Offset the Z-axis an amount greater than the greatest programmed Z-axis depth.

❑ Refer to **Test-Run Without the Workpiece**. Follow those steps to:

■ Set control to dry-run mode.

■ Set feed rate override to zero.

■ Cycle through the program to verify the tool path and clearances.

■ Edit the program to correct any errors.

■ Repeat procedures as needed to prove program operation.

MEETS THE FOLLOWING (MSSC) PRODUCTION STANDARDS

P1-K3	Set up equipment for the production process.
P1-K3B	Setup meets process specifications of internal and external customers.
P1-K3D	Setup procedures are documented for repeatability.
P1-K3F	Setup meets equipment specifications.
P1-K4	Perform and monitor the process to make the product.
P1-K4B	Manufacturing process cycle time meets customer and business needs.
P1-K5G	Necessary adjustments are performed in a timely manner.
P1-O2D	Skill in setting up, programming, and operating the computerized control process.
P1-O4J	Skill in setting up and testing machines.
P1-O6E	Skill in interpreting route sheets and operation sheets to set up and operate machine.
P2-A11A	Acquire new skills necessary to operate high technology equipment.
P2-A11B	Acquire training in setup and maintenance to improve your worth as an employee.
P4-K6	Document product and process compliance with customer requirements.
P4-K6B	Documentation of compliance is written in the appropriate format and correctly stored.
P4-K6C	Documentation of compliance is forwarded to the proper parties.

Calculating Percentages

An adjustment to the volume or size of a product is often expressed as a percentage. A percentage is a fraction or part of a whole. A percentage is often expressed as the part divided by the whole, multiplied by one hundred:

$$\text{Part} \div \text{Whole} \times 100 = \%$$

A customer may order a product that is not standard. The manufacture of such a product may require adjustments to the machines used to manufacture the product. Knowing how to calculate percentages is critical to making accurate and timely machine adjustments.

Finding the Percentage of a Part of a Whole

Example: A metal bar has a total length of 359 mm. A piece 37 mm long is cut from one end. What percentage of the bar was removed?

1. Find the numerical value of the part that was removed.
   The piece cut from one end of the bar measures 37 mm.

2. Find the numerical value of the whole.
   The bar measured 359 mm before the 37-mm piece was removed.

3. Divide the part removed (37 mm) by the whole (359 mm).
   $37 \div 359 = .103$

4. Multiply the resulting number by 100 to yield a percentage value.
   $.103 \times 100 = 10.3\%$

Finding the Percentage to Add to a Whole

Example: A 100-gallon quantity of soft drink syrup is to be increased by 25%. How many gallons of soft drink syrup will be added?

1. Determine the numerical value of the whole. The soft drink syrup is 100 gallons.

2. Determine the percentage to increase the whole.
   The volume of the syrup is to be increased by 25%.

3. Determine the multiplier.
   Divide the percentage value by 100 to obtain the multiplying factor
   $25 \div 100 = .25$
   or move the decimal two places to the left
   $25.0 \rightarrow 0.25$

4. Multiply the whole by the multiplier.
   100 gallons $\times$ .25 = 25 gallons

5. Add the resulting number to the whole.
   100 gallons + 25 gallons = 125 gallons

Finding the Percentage to Remove from a Whole

Example: You need to decrease 100 gallons of soft drink syrup by 33%. How many gallons are represented by this percentage?

1. Determine the numerical value of the whole. There are 100 gallons of soft drink syrup.

2. Determine the percentage to decrease the whole.
   You are decreasing the volume by 33%.

3. Determine the multiplier.
   Divide the percentage value by 100 to obtain the multiplying factor
   $33 \div 100 = .33$
   or move the decimal two places to the left
   $33.0 \rightarrow 0.33$

4. Multiply the whole by the multiplier.
   100 gallons $\times$ .33 = 33 gallons

5. Subtract the resulting number from the whole.
   100 gallons − 33 gallons = 67 gallons

MEETS THE FOLLOWING (MSSC) PRODUCTION STANDARD

P1-A16B	Calculate percentages in order to make machine adjustments.

COPYRIGHT © GLENCOE/McGRAW-HILL

Manufacturing with Metals

Physical properties are those that relate to the interaction of materials with various forms of energy and with other forms of matter. Physical properties can be measured without changing the composition of the material. Mechanical properties are those that are altered when a force is applied to the material. This usually involves changing or destroying the material.

Using the Internet or print resources, research the following physical and mechanical properties of metals listed in the first column of the following table. In the second column, identify whether the property is physical or mechanical. In the third column, identify a type of manufactured product that requires that specific property.

Property	Type of Property	Type of Product Manufactured
compression		
creep strength		
ductility		
elasticity		
electrical conductivity		
hardness		
impact strength		
plasticity		
shear strength		
tensile strength		
thermal expansion		
torsion		

MEETS THE FOLLOWING MSSC PRODUCTION STANDARD

P1-04C	Knowledge of the materials to be used.

Manufacturing with Ceramics

Using the Internet or print resources, research the ceramic materials listed in the first column of the following table. In the second column, identify at least one product or type of product that is manufactured from each specific material. In the third column, identify the properties of this ceramic material that make it ideal for manufacturing the product.

Ceramic Material	Product or Type of Product	Properties
alumina		
beryllium oxide		
boron carbide		
boron nitride		
carbons		
ceramic fibers		
glass ceramics		
graphite		
porcelain		
silica		
silicate glass		
silicon carbide		
silicon nitride		
tungsten carbide		
zirconia		

MEETS THE FOLLOWING (MSSC) PRODUCTION STANDARD

P1-04C	Knowledge of the materials to be used.

COPYRIGHT © GLENCOE/MCGRAW-HILL

Manufacturing with Composites

A composite is a combination of two or more distinct types of materials to create a new, superior material that has properties the original materials do not have. There are a number of types of composites. Using the Internet or print resources, research the composite types listed in the first column of the following table. In the second column, identify an example of the type. In the third column, briefly explain the advantages of this type of composite for use in manufacturing products.

Type of Composite	Example	Advantages in Manufacturing
fiber-reinforced		
laminar		
sandwich		
particulate		
flake		
filled		
hybrid		
smart		

MEETS THE FOLLOWING (MSSC) PRODUCTION STANDARD

P1-04C	Knowledge of the materials to be used.

COPYRIGHT © GLENCOE/MCGRAW-HILL

Manufacturing with Plastics

Many plastics are made from chemicals derived from fossil fuels or petroleum products. Using the Internet or print resources, research the plastics listed in the first column of the following table. In the second column, identify the starting material that is used to make the plastic. In the third column, identify products or types of products that are manufactured from the plastic.

Plastic	Starting Material	Products or Types of Product
acrylic		
epoxy		
fluorocarbon		
phenolic		
polyamide		
polycarbonate		
polyester		
polyethylene		
polypropylene		
polystyrene		
polyurethane		
polyvinyl chloride		
silicone		
unsaturated polyester		

MEETS THE FOLLOWING (MSSC) PRODUCTION STANDARD

P1-04C	Knowledge of the materials to be used.

COPYRIGHT © GLENCOE/McGRAW-HILL

Preventing Metal Corrosion

One weakness of metal as a production material is its vulnerability to corrosion. Corrosion is a destructive electrochemical process that alters and weakens metal. Two causes of corrosion are oxidation and galvanic action.

Oxidation

Oxidation occurs when a metal exposed to air, water, or acid is converted to one of its compounds. For example, the corrosion of ferrous metals (metals containing iron) by oxygen and water produces iron oxides, or rust.

In manufactured products, oxidation is prevented by:

- Covering the metal with paint or oil to keep the air from coming in contact with it.
- Plating it with a compatible metal through chemical solutions or the electroplating process.
- Blending or alloying it with other metals. For instance, alloy steels are carbon steels that are alloyed with other metals. Stainless steel includes chromium, an element that makes the steel more corrosion resistant.

Galvanic Action

When two different metals or alloys are in contact with each other in an electrolyte (such as water), an electric current flows between the two. This galvanic action causes corrosion, i.e., material is removed from one of the metals or alloys (the anode). The other metal (the cathode) is protected. The galvanic series is a guide that lists metals according to their risk of galvanic corrosion. The further apart metals are on the list, the more likely that joining them will result in corrosion of the anodic metal. See p. 128.

For example, a ship's hull is made of steel but has many bronze fittings. Steel is closer to the anodic end of the galvanic series; bronze is further from that end. In seawater, the bronze fittings would attack the steel hull, which would result in corrosion of the steel. To protect the hull, shipbuilders attach zinc plates to the steel. The bronze attacks the zinc, which is more anodic than steel.

Another factor that determines corrosion is the relative size of the anode and the cathode. If the anode is relatively much smaller than the cathode, the corrosion will be more extensive. If the anode is relatively larger than the cathode, the corrosion will be less severe. The temperature and concentration of the electrolyte also determine the corrosion rate.

In production, using one type of metal as a fastener to join pieces of a different metal can cause corrosion. In this case, it's important to combine a large anode with a small cathode. Therefore, fasteners such as bolts and screws should be made of the more cathodic metal to prevent or lessen the chance of corrosion. For example, a steel bolt (small cathode) in an aluminum plate (large anode) is less susceptible to attack because the anode is so large compared to the cathode.

Another way to prevent corrosion is to stop the electrical current from flowing between the two metals by insulating them. For instance, when using a steel bolt in a copper plate, you could use a nonmetal bushing and a nonmetal washer.

(continued on next page)

Galvanic Series

Most active or anodic (easy to corrode)
Magnesium and its alloys
Zinc
Commercially pure aluminum
Cadmium
2024 aluminum
Mild steel or wrought iron
Cast iron
Lead-tin solders
Lead
Tin
Nickel (active)
Hastelloy B (active)
Brasses
Copper
Bronzes
Monel
Silver solder
Nickel (passive)
18-8 stainless steel (passive)
Hastelloy C
Silver
Titanium
Graphite
Gold
Platinum
Most passive or cathodic (resistant to corrosion)

APPLYING YOUR KNOWLEDGE

1. Why is stainless steel resistant to rust?

2. Why is magnesium to steel a bad combination?

3. Why should bolts, screws, and other fasteners be made of a metal less likely to be oxidized than the metal being joined?

PRODUCTION CHALLENGE

You work in quality control for a manufacturer of metal parts used in the assembly of housings for electrical equipment. Customers report that the parts have started to corrode after six months. You note that all of these reports are coming from a specific area of the country. What would you do to identify possible causes of the corrosion?

MEETS THE FOLLOWING (MSSC) PRODUCTION STANDARD

| P1-A17B | Knowledge of metallurgy in order to ensure that corrosive metals are not combined. |

COPYRIGHT © GLENCOE/McGRAW-HILL

Testing Finish Paint Hardness

Painting is an important part of the manufacturing process in a number of subindustries, including transportation equipment, furniture, wood products, and fabricated metal products. In addition to improving appearance, the purpose of paint is to protect surfaces from damage. Besides protecting products, paint is used within the manufacturing facility to protect walls, floors, and equipment from damage and wear. Damage can be caused by:

- Physical contact.
- Chemical reactions with oxygen and water in the air.
- Microorganisms such as mold and bacteria.

Chemical Reactions

Paints contain a mixture of chemicals suspended or dissolved in water, alkyd resin, or oil-based solvents. When the paint dries on a surface, chemical reactions form a hard, uniform, protective coating. The physical properties of paint vary greatly, depending on what materials are used to make the paint. In addition, two coatings made with the same paint can have different properties if the drying conditions, such as temperature and humidity, are different. These factors affect the chemical reactions that form the paint film.

Paint Hardness

One measure of the ability of paint to protect a surface is its hardness, the ability of the film to resist indentation. Coating-hardness testers use indentation, scratching, or rubbing tests to evaluate the hardness or wear resistance of thin films of paint, sealants. Essentially, hardness of a finish paint coating is determined by measuring the amount of force required to cause a permanent dent or scratch in the surface of the coating. Although there are a number of instruments designed to measure paint hardness, it is often tested by a very simple process, using pencil lead to mar the surface.

Film Hardness-Pencil Test

This test method is especially useful during product development and in production control testing in a single work site. The equipment in this test consists of a series of pencils of specified variations in hardness, and a special pencil sharpener for reproducing exactly the same point. Because the hardness of pencils can vary from one manufacturer to another, this test should only be performed using a pencil set specifically designed for the procedure. Sharpening must be done with care, following the test kit manufacturer's instructions exactly in order to ensure reproducible results.

With each test, the pencil is pressed point first against the painted surface, or a score line in the paint, until the pencil lead breaks, or the point penetrates the paint or causes the paint to be released from the surface.

❑ The process is started with the hardest pencil and continued down the scale of hardness to either of two end points:
- One, the pencil that will not cut into or gouge the film (pencil hardness).
- Two, the pencil that will not scratch the film (scratch hardness).

❑ A coated panel is placed on a firm horizontal surface.

❑ The pencil is held firmly against the film at a 45° angle, point away from the operator, and pushed away from the operator in a 0.256-in. (6.5-mm) stroke.

The test can vary from one operator to another, so training is necessary to ensure a standard procedure within the work group.

(continued on next page)

APPLYING YOUR KNOWLEDGE

1. What are the main reasons for painting a surface?

2. Why is it important to control the conditions in the paint drying area when you apply paint to a product, such as an automotive part?

3. What property of the paint coating is measured by the pencil hardness test?

4. Why are the results of this test more likely to vary from one person to another than do the results of many other kinds of tests?

PRODUCTION CHALLENGE

The pencil test is just one way to test paint hardness. More high-tech tests are also used. Part of the quality control of lawn mower production includes using ultrasonic testing to expose the painted parts to corrosive action that simulates continuous exposure to moisture, grass, and other materials in the natural environment. What would you need to know about the chemical properties of the paint that protects the shell of the mower to ensure that it provides superior protection?

MEETS THE FOLLOWING (MSSC) PRODUCTION STANDARD

| P8-A17B | Knowledge of chemical properties to determine if finish paint meets hardness specifications. |

COPYRIGHT © GLENCOE/MCGRAW-HILL

Reading Warning Labels

Occupational Safety and Health Administration (OSHA) standard 1910.1200 mandates that employees have a right to know about the specific hazards of their workplace. Employers are required to develop, implement, and explain a system of communicating such hazards to their employees. This system of hazardous communication is called HAZCOM.

There is no federal standard for the format of the warning labels on hazardous materials. OSHA does not endorse a specific system. Each employer must have a system for HAZCOM, but not all manufacturers use the same system. Consequently, you should be familiar with the specific HAZCOM system present in your workplace. One of the most common means of communication is the Material Safety Data Sheet (MSDS). This is a document that details the specifics of a material, including all potential hazards and precautions.

- All hazardous materials must be clearly and distinctly labeled for easy identification.
- Each hazardous material must also have a corresponding MSDS.
- The MSDS must be available to the employee.

The Department of Transportation (DOT) has a system for identifying hazardous materials during shipment.

- Hazards can be identified by reading codes on the shipping documents or packaging.
- Placards presenting similar information are also used on vehicles to identify the specific hazards of the cargo they are carrying.
- More information on this system and its coding is available in the DOT's *Emergency Response Guidebook*. The *Guidebook* can be downloaded from the DOT's Web site or by contacting the DOT.

The National Paint & Coatings Association's Hazardous Material Identification System (HMIS) has a HAZCOM system that uses colored bars.

- This system employs a label consisting of four bars of different colors.
- These bars indicate the type of hazard and the needed personal protective equipment (PPE).

- An alphanumeric code in each bar identifies the severity of the hazard or the specific PPE needed.

The National Fire Protection Association (NFPA) uses the "hazard diamond" in its HAZCOM system.

- This diamond consists of four colored squares corresponding to the type of hazard.
- A number or symbol in each square corresponds to the severity of the hazard.
- The hazard diamond is intended for use primarily by emergency crews. It contains information that relates largely to the flammability or water reactivity of any substances in a given area.
- This diamond is found mainly on entryway doors to areas where hazardous materials may be stored.

Competency Checklist

As can be seen, HAZCOM systems vary. It is critical that all employees are trained in the recognition of hazards using the HAZCOM system that is in place. The following general checklist will help ensure that all employees are competent in the use of the HAZCOM system.

❑ Regularly review OSHA standards to ensure that the workplace is in compliance with any changes that might have been made since the last update.

❑ Regularly review the current HAZCOM plan to ensure that it is adequate for the work environment.

❑ Regularly review employee training records in the HAZCOM system to make sure that all employees have received training.

❑ Regularly test employee knowledge of the HAZCOM system.

(continued on next page)

☐ Conduct refresher courses that emphasize individual attention for those employees whose knowledge of the HAZCOM system might have lapsed.

☐ Conduct emergency drills to ensure that employees knowledgeable about the HAZCOM system can apply that knowledge in an emergency.

☐ Update the HAZCOM system as appropriate to reflect the most current available format.

Following are examples of some of the warning labels and symbols displayed in the workplace.

MEETS THE FOLLOWING (MSSC) PRODUCTION STANDARDS	
P3-03F	Knowledge of OSHA and other health and safety requirements as applied to the workplace.
P3-A15A	Read warning labels to identify potentially hazardous materials.
P7-K3C	Hazardous materials procedures and policies, such as Material Safety Data Sheet and "right to know", are accurately followed.

COPYRIGHT © GLENCOE/MCGRAW-HILL

Calibrating Air Test Pumps

If you work in an area where there is a possibility of hazardous vapors or particles in the air, you might wear an air quality monitoring device. This device has a small pump that pulls air through a tube or filter that then traps the contaminant. The contents of the tube or filter are analyzed to detect changes in air quality.

Measuring Air Volume

The volume of air drawn through the collection media must be accurately measured. It is necessary to know the air volume in order to determine the concentration of the contaminant being measured. Air pumps have a setting for flow rate and may have a built-in meter to show air flow. To ensure that the flow readings as shown on the pump are accurate, the air pump must be calibrated. The calibration schedule will be based on standards provided by the pump manufacturer or standard operating procedures provided by the company industrial hygiene officer.

How It Works

Pump calibration is done by drawing air through a glass or plastic tube that is marked with lines calibrated to labeled volumes in the tube. A soap solution in a rubber bulb is used to make a thin soap bubble disc across the inside of the tube. As the air flow carries the bubble past the calibration marks, time is measured with a stop watch. The flow rate is calculated from the volume of air required to move the bubble (from the calibration marks) and the number of seconds that it takes to pump that amount of air.

Calibrations can be performed manually or using an electronic calibrator. The electronic system uses a beam of light and a detector to monitor the bubble and an internal clock instead of a stopwatch. These units are high-accuracy electronic bubble flow meters that provide instantaneous airflow readings and cumulative averaging of multiple samples. These calibrators measure the flow rate of gases and present the results as volume per unit of time and should be used to calibrate all air-sampling pumps. Flow rate can be read directly from the meter.

Manual Calibration of Air Pump

Only trained personnel can conduct this calibration test.

1. Allow the pump to run 5 minutes prior to calibration and then shut it off.

2. Assemble the filter holder using the appropriate filter for the sampling method.

3. Connect the collection device, tubing, and pump to the fitting on the calibration device.

4. Inspect all tubing connections to make sure they are airtight.

5. Wet the inside of the calibrated tube with soap solution.

6. Turn on the pump and adjust the pump rotameter to the appropriate flow rate setting.

7. Squeeze the rubber bulb briefly to capture a soap bubble in the tube.

8. Draw two or three bubbles up the tube in order to ensure that at least one bubble will complete its run.

9. Watch a single bubble and time the bubble from 0 to 1000 ml marks for high-flow pumps or 0 to 100 ml marks for low-flow pumps. Calculate the flow rate, in ml/sec by dividing the volume by the time and in liters/min by multiplying the calculated rate by 0.060.

10. The timing accuracy must be within ±1 second of the time corresponding to the desired flow rate. If the time is not within the range of accuracy, adjust the flow rate and repeat steps 8 and 9 until the correct flow rate is achieved. Perform steps 8 and 9 at least twice.

11. Mark on the calibration sheet or pump label the date and the calibrated flow rate.

(continued on next page)

Electronic Bubble Meter
Calibration of Air Pump

Follow the procedure for manual calibration, substituting the following for steps 7–10. Only trained personnel can conduct this calibration test.

7. Press the button on the electronic bubble meter.

8. Electronically time the bubble. The accompanying printer will automatically record the calibration reading in liters per minute.

9. Repeat step 8 until two readings are within 2%.

10. If necessary, adjust the pump while it is still running.

MEETS THE FOLLOWING (MSSC) PRODUCTION STANDARD

P4-A17D   Conduct air quality test equipment calibration to ensure worker safety.

COPYRIGHT © GLENCOE/McGRAW-HILL

Locating & Drilling Holes

Milling machines are versatile tools that can be used to perform a wide variety of operations. Some milling machines resemble a conventional drill press. These are known as vertical milling machines. The following is an example of an operation that can be performed on such a machine. Before operating a milling machine, you must know how to make several kinds of adjustments on the machine. These adjustments include those made to the knee, the table, the speed, and the feed.

Safety
- Make sure you are wearing the appropriate personal protective equipment, including eye protection.
- Observe all safety precautions when operating this tool.

Knee Adjustments
- This is raised or lowered to establish the proper elevation of the workpiece under the cutting tool.
- It is generally best to establish the depth of cut by moving the knee, though this is not necessary on a vertical mill.

Transverse Table Movement
- The movement of the saddle and table is toward or away from the column.
- The saddle is clamped to the knee for most milling operations to dampen vibration.
- The saddle must be loosened before making movements.
- The table is moved by turning the cross-feed hand wheel or handcrank.

Longitudinal Table Movement
- The movement of the table is from side to side (left or right).
- The table is moved by turning the table transverse hand wheel or longitudinal feed.
- Most machines are equipped with power longitudinal feed.
- Most machines are equipped with a table clamp lever that should be tightened during hole machining operations.

Rapid Traverse Control
- Allows the operator to move the knee or table in any direction with power.

Spindle Control
- Spindle speed is designated in rpm (revolutions per minute).
- Speed can be changed in several ways, including use of:
  - A speed change dial.
  - Levers that shift gears to the desired rpm.
  - A variable speed drive while the machine is running.
  - A step pulley and V-belt drive by shifting belts to different steps with the motor off.

Feed Control
- The feed rate is regulated through a series of change gears in a feed change gearbox.
- Sometimes a feed dial is available that can be turned to the exact rate.
- In machines without a power feed, the rate must be controlled directly by the operator.
- Feeding too rapidly can lead to tool breakage.
- Feeding too slowly can lead to excess tool wear.

(continued on next page)

Locating and drilling a hole in a precise location using a vertical mill.

Consult the equipment operations manual before undertaking this operation. These are general procedures. Your manual may present different instructions. Proper setup is critical to producing a quality product. Follow the prescribed procedures for setup when performing any kind of machining operation. Conduct a thorough inspection of the setup to affirm that everything is correct before proceeding with the operation. Occasionally, it may be necessary to make on-process adjustments during production. If that is the case, many of the procedures used in setting up the process can be used to adjust it during production.

1. Install a drill chuck in the spindle.

2. Install a dial indicator in the drill chuck.

3. Check the vise alignment and the vertical position of the spindle.

4. Mount the workpiece in the vise so that holes will not be drilled into the vise bottom or supporting parallels.

5. Install an edge finder in the drill chuck.

6. Position the end of the edge finder about ¼" (6.35 mm) below the top surface of the workpiece.

7. Set the spindle speed at about 600 rpm.

8. Release the micrometer dials on the hand wheels.

9. Start the machine and move the table in the longitudinal direction until the end of the workpiece contacts the side of the edge finder and makes it jump off center.

10. Move the edge finder above the top surface of the workpiece.

11. Set the longitudinal micrometer dial for 0.

12. Move the spindle half of the edge finder diameter (usually 0.250" [6.35 mm]) plus the distance the center of the hole is located from the end of the workpiece.

13. Set the longitudinal table lock.

14. Move the workpiece in the transverse direction far enough to position the end of the edge finder ¼" (6.35 mm) below the top surface of the workpiece.

15. Repeat steps 9 through 12 to locate the center of the hole in the transverse direction.

16. Set the transverse table lock and stop the machine.

17. Remove the edge finder from the drill chuck.

18. Install a center drill in the drill chuck.

19. Drill a center hole.

20. Follow the center drill with drills and other drilling tools as required. To provide room for tool changing, raise the spindle or lower the table. Do not move the workpiece longitudinally or transversely.

MEETS THE FOLLOWING (MSSC) PRODUCTION STANDARDS

P2-01L	Skill in making on-process adjustments during production.
P4-01A	Skill in setup and inspection to improve production and maintain quality.

COPYRIGHT © GLENCOE/MCGRAW-HILL

Performing Precision Turning

The lathe is a common tool in the manufacturing industry because it can be used to produce a wide variety of products that require precision turning. Each lathe is different. It will be necessary to discuss the operation of a specific lathe with a supervisor. To operate a lathe correctly, you must first become familiar with its parts.

- The lathe bed is the main frame upon which the machine is built.
- The headstock is permanently fastened to the left end of the lathe bed.
- The tailstock can be clamped at any point along the lathe bed to the right of the headstock.
- Both the headstock and the tailstock have spindles.
- The headstock spindle is driven by a motor in a cabinet beneath the headstock.
- The tailstock spindle does not turn. However, it can be moved in and out of the tailstock with the tailstock hand wheel.
- Both spindles can hold lathe centers that support long workpieces between them.
- The tapered hole of the tailstock spindle also accepts drill chucks and tapered-shank drills and reamers for drilling operations.
- The carriage is the part that slides back and forth on the lathe bed between the headstock and the tailstock and carries the cutting tool.
- The precision V-shaped and flat surfaces of the bed, on which the carriage and tailstock slide, are called bed ways.

Safety
- Make sure you are wearing the appropriate personal protective equipment, including eye protection.
- Observe all safety precautions when operating this tool.

Before beginning any process involving tools or machinery, follow these procedures.

- Check with your supervisor to ensure that resources such as materials, tools, and equipment are available for the production process.
- Check to ensure that all necessary resources are at the workstation.
- Check that the setup for any machine process meets all relevant ergonomic, health, safety, and environmental standards.
- Check to make sure that proper repairs and adjustments have been made.

Turning a Precision Diameter

Consult the equipment operations manual before undertaking this operation. These are general procedures. Your manual may present different instructions.

1. Mount the workpiece accurately and securely.
2. Mount the tool holder and cutting tool and position them properly.
3. Set the lathe for the correct rpm and the gearbox for the desired feed rate.
4. The reference cut is the first cut, from which other cutting measurements start. To start this cut, turn on the lathe and advance the cutting tool until it touches the workpiece.
5. Then move the tool off to the right end of the workpiece.
6. Next, advance the tool toward the center of the workpiece for an additional amount that leaves the workpiece larger in diameter than the desired finish size.

(continued on next page)

7. Be sure to leave 0.010" to 0.030" (0.25 mm to 0.76 mm) on the diameter for a finish cut. Turn a cylinder about ⅛" (3.2 mm) long at this setting.

8. Stop the lathe and measure the diameter obtained.

9. Reposition the tool, if necessary, before completing the rough cut.

10. Stop the lathe.

11. Move the tool off the right end of the workpiece, and advance it far enough to make the cylinder the correct diameter.

12. Start the lathe and allow the cut to proceed about ⅛" (3.2 mm).

13. Stop the lathe and measure the diameter.

14. Adjust the tool position if necessary and make the finish cut.

The lathe accomplishes this cutting task by employing some basic mechanical principles. Keep these principles in mind when using a lathe:

- The workpiece itself rotates while the cutting tool remains stationary.
- The torque supplied by the motor to the workpiece ensures that the workpiece is constantly moving at the speed necessary for cutting to be performed.
- The cutting tool itself is essentially a wedge.
- When this wedge is forced against a workpiece moving at speed, it breaks the surface of the object and begins peeling or chipping it off in thin layers.
- To facilitate this process and diminish the heat buildup from friction, a cutting fluid is often applied.
- The cutting fluid acts as a lubricant that makes the cutting tool more efficient at removing metal.
- The cutting fluid also acts as a coolant that helps reduce heat buildup from friction.

MEETS THE FOLLOWING (MSSC) PRODUCTION STANDARDS

P1-K2	Determine that resources such as materials, tools and equipment are available for the production process.
P1-K2D	Necessary resources are at workstation when required.
P1-K3A	Proper repairs and adjustments are made to production equipment prior to putting into service.
P1-K3E	Setup meets ergonomic and other relevant health, safety, and environmental standards.
P1-O2E	Skill in operating production equipment.
P1-O6B	Skill in making machine adjustments.
P2-A11D	Learn new skills related to all parts of the production process.
P2-A17A	Understand the mechanical principles of machinery.

COPYRIGHT © GLENCOE/MCGRAW-HILL

Boring with a Vertical Mill

Safety and efficiency are the hallmarks of high-performance manufacturing. To achieve maximum safety and maximum efficiency, production workers must be skilled in the operation of different machines. Consider the following situation. A machine tool malfunctions in a manner that creates a potential safety hazard. The machine will need to be repaired. To avoid the production shutdown that would otherwise have occurred, another machine will be used to perform the operations of the malfunctioning machine tool.

Production equipment breaks down, job requirements change, safer outcomes might be required, or temporary safety issues might arise that could potentially halt production. Training operators to perform similar jobs on different machines and updating that training regularly offers a means to overcome obstacles that might halt production. It helps ensure that production workflow runs efficiently and that production schedules are met.

With cylindrical parts, boring operations are typically performed on the lathe. However, with noncylindrical parts, if the lathe breaks down or job requirements change, it is possible to use a vertical milling machine to compensate for the unexpected breakdown of a lathe. The general procedure for boring on a vertical milling machine follows.

Boring on a Vertical Milling Machine

Safety
- Make sure you are wearing the appropriate personal protective equipment, including eye protection.
- Observe all safety precautions when operating this tool.

Consult the equipment operations manual before undertaking this operation. These are general procedures. Your manual may present different instructions.

1. Install a drill chuck in the spindle.
2. Install a dial indicator in the drill chuck.
3. Check the vise alignment and the vertical position of the spindle.
4. Mount the workpiece in the vise so that holes will not be drilled into the vise bottom or supporting parallels.
5. Install an edge finder in the drill chuck.
6. Position the end of the edge finder about $\frac{1}{4}$" (6.35 mm) below the top surface of the workpiece.
7. Set the spindle speed at about 600 rpm.
8. Release the micrometer dials on the hand wheels.
9. Start the machine and move the table in the longitudinal direction until the end of the workpiece contacts the side of the edge finder and makes it jump off center.
10. Move the edge finder above the top surface of the workpiece.
11. Set the longitudinal micrometer dial for 0.
12. Move the spindle half of the edge finder diameter (usually $\frac{1}{4}$" [6.35 mm]) plus the distance the center of the hole is located from the end of the workpiece.
13. Set the longitudinal table lock.
14. Move the workpiece in the transverse direction far enough to position the end of the edge finder $\frac{1}{4}$" (6.35 mm) below the top surface of the workpiece.
15. Repeat Steps 9 through 12 to locate the center of the hole in the transverse direction.
16. Set the transverse table lock and stop the machine.
17. Remove the edge finder from the drill chuck.
18. Install a center drill in the drill chuck.
19. Drill a center hole.

(continued on next page)

20. Follow the center drill with other drilling tools to create a hole that is about $1/16$" (1.6 mm) smaller than the hole required. To provide room for tool changing, raise the spindle or lower the table. Do not move the workpiece longitudinally or transversely.

21. Install a boring head.

22. Install a boring tool of suitable size in the boring head.

23. Adjust the boring head until the boring tool touches the side of the drilled hole. Withdraw the boring tool from the hole and adjust the boring head to enlarge the diameter by about 0.020" (0.5 mm).

24. Start the machine, engage the power feed, and bore the hole about $1/8$" (3.2 mm) deep. Stop the machine, withdraw the boring tool, and measure the hole diameter.

25. If the hole is not oversize, start the machine and finish boring the hole.

26. Stop the machine, withdraw the boring bar, and measure the hole.

27. Adjust the boring head for the finish cut, start the machine, and bore the hole about $1/8$" (3.2 mm) deep. Then stop the machine and measure the hole.

28. Adjust the boring head, if necessary, to make the hole the correct diameter. Then continue with the finishing cut.

MEETS THE FOLLOWING (MSSC) PRODUCTION STANDARDS

P2-A7A	Revise workstation equipment to meet new job requirements.
P2-A7C	Use a different machine tool to compensate for an unexpected tool breakage.
P3-A7A	Change method of production to achieve safer outcomes.
P3-A7C	Change the production process to temporarily work around an unsafe area or condition.
P6-K4A	Production schedules are met effectively.
P6-K4C	Production workflow runs efficiently.

COPYRIGHT © GLENCOE/McGRAW-HILL

Soldering Basics

Soldering is the joining of metallic surfaces by a molten metallic alloy. Such an alloy has a relatively low melting point. This alloy is known as *solder*. Many metals and their alloys can be used as solders. Solder is available in various forms and thicknesses.

Soldering is performed using a soldering iron or a soldering gun. Both soldering irons and soldering guns are electrical devices. Soldering guns generally have a higher wattage than soldering irons. Soldering requires less skill on the part of the operator than brazing or welding. The following are general steps. They may vary, depending on the item being soldered.

> **Safety Note:** The rosin in solder releases fumes that are harmful to eyes and lungs.
> - Always solder in a well-ventilated area.
> - Wear eye protection.
> - Wear cotton gloves while soldering. Be careful when working with hot solder. It will burn you.
> - Wash your hands after soldering.

1. Clean all surfaces to be soldered. The surfaces to be soldered must be free of dirt, grease, and nonmetallic matter. Depending on the materials to be soldered, you can use steel wool or a steel-bristle brush. Solvents such as alcohol can also be used. Clean surfaces will help ensure a strong joint.

2. If possible, secure the workpiece so that it does not shift.

3. Select the proper soldering iron. It will take longer to heat up a large joint with a low-wattage soldering iron than with an iron of greater wattage.

4. Solder small parts before large parts. This will make assembly easier.

5. Operate the soldering iron at the lowest possible temperature.

6. Place a small amount of solder on the hot iron tip of the soldering iron.

7. Wipe the tip of the soldering iron on a damp sponge. Leave a thin layer of solder on the iron.

8. Hold the surfaces to be joined slightly apart.

9. Have ready solder of adequate diameter.

10. Do not add more solder than is needed. It is important to use the right amount of solder on the iron and the joint. Excess solder will create a poor joint.

11. Place the tip of the soldering iron so that it contacts both of the surfaces to be joined. It is important that both parts to be joined are at the same high temperature.

12. Melt a small amount of solder onto the tip of the soldering iron at the point where the soldering iron is touching both parts of the joint. The solder will melt.

13. Place suitable solder in or near the joint. Allow just enough solder to fill the space.

14. Apply heat until the solder flows into the joint. The application of the solder should take no more than a few seconds.

15. Remove the solder first. Then remove the tip of the soldering iron. Be careful not to allow the joint connection to move while the solder is solidifying.

16. Do not move the parts until the solder is completely hard. A joint that is disturbed during this cooling period may become seriously weakened.

(continued on next page)

17. Assess the quality of the joint.
    - A properly soldered joint should be bright, smooth, and shiny.
    - A joint that is rough or one that has ridges or sharp points was probably overheated. It must be redone.
    - A joint that has a grainy appearance was either moved before it cooled or was not heated adequately. This is known as a cold joint. It is caused by insufficient heat or by foreign material in the connection. It must be redone.

18. To redo a joint, reheat the joint and apply a small amount of solder. If the added solder would place too much solder on the joint, the solder will need to be removed and the joint resoldered.

19. Keep the soldering iron tip clean. A clean tip will conduct heat more effectively.

20. If necessary, finish the exposed solder to match the surrounding surface. Exposed solder may be ground, filed, polished, painted, or electroplated.

MEETS THE FOLLOWING PRODUCTION STANDARD

P1-A17C	Understand the soldering process and how it works.

COPYRIGHT © GLENCOE/MCGRAW-HILL

Handling Chemical Spills

Accidents resulting in the release of chemicals will occur despite the best effort to work safely. Prior to working with any chemical, you must be familiar with its hazards and proper handling procedures. This information is available in the company safety manual and in the Material Safety Data Sheet (MSDS) that is on file for each chemical handled on the site. The person who causes or discovers a spill is responsible for ensuring that proper procedures are followed so that the spill is cleaned up in a manner that protects personnel and the environment. If you are not sure of procedures, notify your supervisor.

Evaluating the Spill

Before working with any chemical, review the MSDS and other available safety information to become familiar with any hazards. If a spill occurs, determine whether it is a minor spill or a major spill.

Minor Chemical Spills. A minor spill is one that, due to the type or amount of chemical, does not pose a risk to health and does not have the potential to become an emergency. Follow these steps to clean up a minor spill.

1. Notify other personnel of the accident.
2. Decide whether you can safely handle the spill. Consult MSDS and other safety documents to determine correct procedures. If you are unsure whether you can handle the spill, notify the Emergency Response Team.
3. Remove all ignition sources if the spill involves flammable materials.
4. Provide ventilation if the spill involves vapors in an enclosed area.
5. Confine the spill to a small area so that it does not spread. Contain the size of a liquid spill with appropriate absorbing materials, such as vermiculite, commercial absorbents, or spill sock (long fabric tube filled with absorbent material).
6. Choose appropriate personal protective equipment (PPE), such as goggles, face shield, impervious gloves, lab coat, and respirator. *Note: All personnel must have medical approval and be fit tested before using a respirator.*

7. Clean up the spilled material and place it in an appropriate disposal container.
   - Sweep solid material into a dustpan and place in a sealed plastic container.
   - Absorb liquids into absorbent material and place in disposal container.
8. Decontaminate the area with soap and water after cleanup and place residue in the proper disposal container.
9. Label all disposal containers with appropriate labels and discard following waste disposal procedures.
10. Notify your supervisor of the spill and fill out any necessary documentation.

Major Chemical Spills. If you are not certain of the identity or hazards of the spilled material, the incident must be considered a major spill. A major spill is any spill that:

- Poses an immediate risk to health.
- Poses a risk of uncontrolled fire or explosion.
- Involves personal injury or risk of personal injury.

Follow these steps to clean up a major spill:

1. Alert personnel in the affected and nearby areas. Evacuate the area immediately if there is a danger of injury. If possible, use a sign or barricade to warn others.
2. Notify the Emergency Response Team or call 911, depending on company safety procedures.

(continued on next page)

3. Attend to injured or contaminated personnel if you can do so without risk to your own safety. Otherwise, be prepared to direct emergency personnel to their location.

4. Close doors to affected areas once the areas are completely evacuated.

5. Remain upwind of any spill that involves hazardous vapors. Evacuate to a designated evacuation assembly area.

6. When emergency personnel arrive, provide as much information as possible about the nature and extent of the spill and be ready to assist as needed.

7. Track and document communication regarding the incident as appropriate.

8. File an incident report, following the procedures in the company safety manual.

MEETS THE FOLLOWING (MSSC) PRODUCTION STANDARDS

P1-07C	Knowledge of safety procedures for chemical spills.
P3-K2B	Emergency response complies with company and regulatory policies and procedures.
P5-K1H	Communications are tracked and documented, as appropriate.
P5-A17A	Knowledge of proper disposal of chemicals.

COPYRIGHT © GLENCOE/McGRAW-HILL

Locating & Using a Material Safety Data Sheet

Chemicals are part of most manufacturing processes. Chemical manufacturers may use hundreds or thousands of different chemical substances. Other manufacturers may use only a few, often for cleaning or finishing processes. The regulations of the Occupational Safety and Health Administration (OSHA) require the label for every hazardous chemical substance to include identification that can be cross-referenced to a Material Safety Data Sheet (MSDS). A MSDS is a document that:

- Contains information on the potential health effects of exposure to the chemical.
- Explains how to work safely with the chemical.
- Identifies hazards in the chemical's use.
- Specifies storage, handling, and emergency procedures.
- States what to expect if recommendations are not followed.
- Explains how to recognize symptoms of overexposure and what to do if overexposure occurs.
- Provides contact information for the chemical supplier.

Your employer is required by law to make the MSDS for every chemical used at the workplace available to all workers. This includes certain types of equipment that contain hazardous chemicals.

Follow these guidelines for using a MSDS:

❑ Know where your facility files MSDS.
- Material Safety Data Sheets should be stored at a central location at the work site. They can be stored as a computer file if all employees have access to the data and training on its use.
- If the MSDS file is on the computer system, there must be a backup system for use in case of a power failure or computer problem. The backup system may consist of filed paper copies or a battery-powered laptop computer separate from the main computer system.

❑ Know how to locate the right MSDS.
- Before starting to work with a hazardous chemical, locate and read its MSDS.

❑ Know how to use the information in the MSDS.
- Remember that even if safety information is provided in the SOP, the MSDS may have additional information.
- In case of a spill or accident, consult the MSDS for first-aid and cleanup procedures. It contains information that you will need to handle the spill as well as information that medical personnel will need to treat an accident victim.
- If you need more information about safe handling of materials, refer to the MSDS for the supplier's contact information.

OSHA's Hazard Communication Standard specifies certain information that must be included in a MSDS. However, it does not require that any particular format be followed in presenting this information. Specified information includes:

- Identity of the hazardous material as it appears on the label.
- The chemical and common name of the hazardous chemical or hazardous ingredients in a mixture.
- Physical and chemical characteristics of the hazardous chemical, such as vapor pressure and flash point.
- Physical hazards of the hazardous chemical, including the potential for fire, explosion, and reactivity.
- Health hazards of the hazardous chemical, including signs and symptoms of exposure, and any medical conditions that are aggravated by exposure.
- The primary route(s) of entry to the body.
- The permissible exposure limit.
- Whether the chemical has been determined to be carcinogenic (cancer causing).

(continued on next page)

- Precautions for safe handling and use, including appropriate hygienic practices, protective measures during repair and maintenance of contaminated equipment, and procedures for cleanup of spills and leaks.

- Control measures such as engineering controls, work practices, or personal protective equipment.

- Emergency and first-aid procedures.

- The date the MSDS was prepared or of the last change to it.

- The name, address, and telephone number of the chemical manufacturer, importer, employer, or other responsible party preparing or distributing the MSDS, who can provide additional information on the hazardous chemical and appropriate emergency procedures, if necessary.

A wide variety of MSDS forms are provided by chemical suppliers. The chemical described in the following sample MSDS is used to manufacture anaerobic adhesives and sealants and pressure-sensitive adhesives. This example MSDS is a shorter version of an actual 16-section MSDS. Some parts are not included. It will guide you when you read actual MSDS documents for hazardous chemicals at your workplace.

MEETS THE FOLLOWING (MSSC) PRODUCTION STANDARDS

P2-04C	Knowledge of how to use and store hazardous materials and chemicals (e.g., compliance with MSDS).
P2-A17B	Knowledge of the chemical with which you are working.
P2-A17C	Understanding of chemicals so as to properly store dangerous materials and chemicals.
P3-01A	Knowledge of how to locate and use Material Safety Data Sheets (MSDS).
P3-A15C	Read MSDS forms to protect self and others.
P3-A17B	Understanding of potential chemical hazards.
P5-03B	Knowledge of how to use Material Safety Data Sheets (MSDS).
P7-01D, P7-04K	Knowledge of Material Safety Data Sheets (MSDS).
P7-A2D	Organize information to meet safety requirements; put safety information at a central location for all maintenance persons involved.
P7-A15C	Review the Material Safety Data Sheets for a new piece of equipment.
P7-A17D	Identify which chemicals are in the facility by Material Safety Data Sheets.
P8-08C	Knowledge of chemicals and the Material Safety Data Sheets (MSDS) used to perform quality checks to ensure safety gear is accessible and present.
CORE-01B	Knowledge of the proper uses of Material Safety Data Sheets (MSDS).
CORE-01C	Knowledge of HAZMAT procedures.

(continued on next page)

   COPYRIGHT © GLENCOE/MCGRAW-HILL

Huntsman Advanced Materials Americas Inc.

HUNTSMAN

281 Fields Lane
Brewster, NY 10509

8am to 4:30pm Phone: (914) 785-3000
24-Hour Health/Environmental Emergency Phone: 1-888-354-3323

| **Effective Date:** 1/29/04 | **Material Safety Data Sheet** | MSDS No: 6893 |

1. PRODUCT IDENTIFICATION

Health	3
Flammability	3
Reactivity	0
Protective Equipment	X

HMIS RATING

Trade Name: Araldite GZ 6060 KX-80

Chemical Family: Epoxide

Intended Use or Product Type: Epoxy Resin Solution

3. HAZARDS IDENTIFICATION

Emergency Overview: Warning! Flammable liquid. Can cause irritation and dermatitis. May cause central nervous system depression.

Primary Route(s) of Entry: Dermal, inhalation.

Carcinogenicity (NTP, IARC, OSHA): Ethylbenzene is considered to be carcinogenic by IARC (Group 2B).

4. FIRST AID MEASURES

Ingestion: If swallowed, give at least 3-4 glasses of water but do not induce vomiting. If vomiting occurs, give water again. Do not give anything by mouth to an unconscious or convulsing person. Get medical attention. Have physician determine whether vomiting or stomach evacuation is necessary.

Skin: For skin contact, wash with large amounts of running water, and soap, if available, for 15 minutes. Remove contaminated clothing and shoes. Get immediate medical attention. Discard or decontaminate clothing before re-use and destroy contaminated shoes.

Inhalation: If inhaled, remove from area to fresh air. If not breathing, give artificial respiration. Get immediate medical attention. If breathing is difficult, transport to medical care and, if available, give supplemental oxygen.

Eyes: For eye contact, immediately flush eyes for at least 15 minutes with running water. Hold eyelids apart to ensure rinsing of the entire eye surface and lids with water. Get immediate medical attention.

Overexposure Effects: Overexposure to this material can cause skin, eye and respiratory irritation. Can cause central nervous system depression, weakness, nausea, lightheadedness, vomiting, dizziness and incoordination. In animal studies, components in this product have caused liver and kidney damage and reproductive effects.

Medical Conditions Aggravated by Exposure: Allergy, eczema or skin conditions.

Additional Information: Immediately remove wet contaminated clothing to avoid flammability hazard. Wash before reuse.

(continued on next page)

Effective Date: 1/29/04	**Araldite GZ 6060 KX-80**	**Huntsman Advanced Materials** **Americas Inc.**

5. FIRE FIGHTING MEASURES

Flash Point: 84°F (29°C)
Flash Point Method Used: Closed Cup

Fire Fighting Extinguishing Media: Carbon dioxide, foam, dry chemical, water spray.

Fire Fighting Equipment: Use self-contained breathing apparatus.

Fire and Explosion Hazards: Decomposition and combustion products may be toxic. Dangerous fire hazard and moderate explosion hazard when exposed to heat and flame.

6. ACCIDENTAL RELEASE MEASURES

Accidental Release Measures: Remove all sources of ignition. Provide explosion proof ventilation. Take up on absorbent material. Shovel with non-sparking tools into closable container for disposal. Wear protective equipment specified below. Thoroughly flush contaminated area with water.

7. HANDLING AND STORAGE

Precautions: Avoid contact with eyes, skin and clothing. Avoid breathing vapor, mist or spray. Use only with good ventilation. Individuals should wash thoroughly after handling. For industrial use only.

Storage Information: Keep away from heat, sparks and open flame. Ground and bond metal containers for liquid transfer to avoid static sparks. Store in closed containers in cool, well-ventilated area.

8. EXPOSURE CONTROLS / PERSONAL PROTECTION

Personal Protective Equipment: Wear protective equipment to avoid personal contact.

Skin Protection: Wear impervious gloves.

Respiratory Protection: Organic chemical cartridge respirator, if needed.

Eye Protection: Wear splash-proof chemical goggles.

Engineering Controls: Explosion-proof general mechanical ventilation. Local exhaust if needed.

9. PHYSICAL AND CHEMICAL PROPERTIES

Color: Clear, Yellow
Odor: Aromatic and Ketone Odors
Physical State: Viscous Liquid
Solubility in Water: Very Slight
Specific Gravity: 9 lb/gal at 25°C (77 °F)
Vapor Density: > Air

Percent Volatile: 20%

COPYRIGHT © GLENCOE/McGRAW-HILL

Storing Tools Correctly

Proper tool and equipment storage affects production in several ways. It leads to efficiency because a tool or piece of equipment that is stored in its assigned place can be quickly retrieved. If properly stored, a tool or item of equipment will not be damaged by conditions of damp or temperature extremes. Tools and items of equipment stored in a secured area cannot be stolen or injure someone. Properly stored tools have a longer service life.

Specific tool and equipment storage guidelines vary by industry. General guidelines follow. Note that proper tool and equipment storage are crucial for:

- Safety and security.
- Protection.
- Order.

Safety & Security

❑ Store powder-actuated tools in a locked container when not being used.

❑ Do not store unguarded cutting tools in drawers. Hand injuries can be caused by searching for such a tool in an assortment of sharp tools.

❑ Store edged tools to protect the blades and bit edges from damage.

❑ Store knives in their scabbards.

❑ Hang saws with the blades away from reach.

❑ Rack heavy tools with the heavy end down.

❑ Ensure that the storage area is properly secured to protect the contents from theft.

Protection

❑ Ensure that the storage area has adequate fire protection.

❑ Ensure that the storage area is dry.

❑ Ensure that the storage area is clean.

❑ Ensure that storage racks, bins, and cabinets are sufficiently strong and in good repair.

❑ Ensure that the storage area is not subject to temperature extremes.

Order

❑ Ensure that the storage area is well lit to allow tools to be easily identified.

❑ Ensure that racks and storage hooks are strong enough to hold the tools.

❑ Return all tools to the correct storage site.

❑ Store each tool and item of equipment in its assigned place. This eliminates searching.

MEETS THE FOLLOWING PRODUCTION STANDARDS

P2-K4A	Tools are stored in proper location.
P2-A9B	Inspire production workers to maintain proper tooling storage in order to eliminate searching.

Evaluating Tool Ergonomics

Ergonomics is the science of designing a task to maximize the efficiency of labor while minimizing effort and energy. In analyzing a task, you must consider not only the materials needed, but also the workplace and the tools. Many job injuries are related to stress and strain caused by tools that are improperly designed or incorrectly used. The general design of the workplace also affects workplace health and safety.

The following checklist is a general survey of how certain work practices related to tools contribute to workplace health and safety. By practicing the science of ergonomics, it is possible to reduce or eliminate certain types of injuries. This will make the workplace safer for everyone. A "No" response indicates potential problem areas that should receive further investigation. Tools and machines may need to be modified to prevent injuries and improve ergonomics. This may require specialized skills. Employee training may be needed. The solution to the problem will depend on the frequency and duration of the activity.

Tool Evaluation Checklist	Yes	No
1. Are tools selected to limit or minimize:		
• exposure to excessive vibration?		
• use of excessive force?		
• bending of the wrist?		
• twisting of the wrist?		
• finger pinch grip?		
• problems associated with trigger finger?		
2. Are powered rather than hand tools used where necessary and feasible?		
3. Are tools evenly balanced?		
4. Are heavy tools suspended in ways that facilitate use?		
5. Are heavy tools counterbalanced in ways that facilitate use?		
6. Does the tool allow adequate visibility of the work?		
7. Does the tool grip/handle prevent slipping during use?		
8. Are tools equipped with handles of textured, nonconductive material?		
9. Are different handle sizes available to fit a wide range of hand sizes?		
10. Is the tool handle designed not to dig into the palm of the hand?		
11. Can the tool be used safely with gloves?		
12. Can the tool be used with either hand?		
13. Is there a preventive maintenance program to keep tools operating as designed?		
14. Have employees been trained		
• in the proper use of tools?		
• when and how to report problems with tools?		
• in proper tool maintenance?		

MEETS THE FOLLOWING (MSSC) PRODUCTION STANDARDS

P3-A17A	Understanding of how the body is impacted by ergonomics in order to make workstation more comfortable and safe.
P7-04M	Skill at modifying machines as prescribed to prevent injuries and improve ergonomics.

COPYRIGHT © GLENCOE/MCGRAW-HILL

Evaluating Workplace Ergonomics

Ergonomics is the science of designing a task to maximize the efficiency of labor while minimizing effort and energy. A task includes the tools with which an operation is performed, the materials on which an operation is performed, and the workspace within which an operation is performed. It is important to have a basic understanding of the ways in which actions such as lifting and pushing can affect your safety and health in the workplace. Reducing or eliminating those actions that draw a "Yes" response on the following checklist can greatly reduce the potential for injury in the workplace. A "Yes" response indicates a potential problem that should be investigated. The solution would depend on the duration and frequency of the action.

Evaluating Motion in the Workplace	Yes	No
Does the lift involve pinching to hold the object?		
Is heavy lifting done with one hand?		
Are very heavy items lifted without the assistance of a mechanical device?		
Are heavy items lifted while bending over, reaching above shoulder height, or twisting?		
Are dollies, pallet jacks, or other carts difficult to get started?		
Could debris (e.g., pieces of broken pallets) or uneven surfaces (e.g., cracks in the floor) or dock plates catch the wheels while pushing?		
Is pulling rather than pushing routinely used to move an object?		
Are heavy objects carried manually for a long distance?		
Are repetitive motions performed for several hours without a break?		
Does the job require repeated finger force?		
Is the back bent or twisted while lifting or holding heavy items?		
Are objects lifted out of or put into cramped spaces?		
Do routine tasks involve leaning, bending forward, kneeling, or squatting?		
Do routine tasks involve working with the wrists in a bent or twisted position?		
Are routine tasks done with the hands below the waist or above the shoulders?		
Are routine tasks done behind or to the sides of the body?		
Does the job require standing for most of the shift without anti-fatigue mats?		
Do employees work with their arms or hands in the same position for long periods of time without changing positions or resting?		
Are there sharp or hard edges with which the worker may come in contact?		

(continued on next page)

Priorities in Workspace Design

Careful attention to the ways in which individuals move in the workplace can influence workplace design. The main priorities of such workplace design are presented in the following list. A careful study of this subject can lead to more detailed priorities.

Priority 1. Anticipate all potential safety hazards and required emergency actions.

Priority 2. Consider the primary visual tasks. Whether it is to tighten a bolt or to monitor a display device, the eye position relative to the task establishes the basic layout of the workplace.

Priority 3. Determine the placement of the primary controls associated with the primary visual task (e.g., operating a power lever). Emergency controls are also primary controls.

- Primary and emergency control positions are sometimes related to where the operator is seated. This is an appropriate time to identify that location.
- Consider the arm reach of the operator when locating controls. Consider also clearance requirements for the operator's head and knees. Be sure to consider the general workforce, not just the "average" person.

- Primary controls requiring precision operation should be positioned for use by either hand. If that is not possible, preference is given to the right hand.
- Emergency controls should be equally available to both hands.

Priority 4. Consider the control/display relationships.

- Controls should be near the displays they affect.
- Activating the control should not obscure the display.

Priority 5. Arrange the workstation elements in their anticipated sequence of operation (usually from left to right and top to bottom). In some cases, this means that similar controls and displays are in the same place for similar types of equipment. This helps eliminate unnecessary retraining.

Priority 6. Position the workstation elements according to frequency of use.

Attention to these priorities can help provide a workspace that is safer for the worker. This will reduce workplace injuries. It will also increase efficiency and job satisfaction. All of these can lead to increased production and profitability.

MEETS THE FOLLOWING PRODUCTION STANDARDS

P3-A17A	Understanding of how the body is impacted by ergonomics in order to make workstation more comfortable and safe.
P7-A17B	Apply knowledge of physics and chemistry to safety activities in the workplace.

                    COPYRIGHT © GLENCOE/McGRAW-HILL

Using Computers in Manufacturing

Computerized manufacturing has brought about many positive changes. Manufacturing is now safer and more efficient. To excel in manufacturing today, it is necessary to have a basic understanding of computers and the software used to manage information. The computer is a valuable tool for storing and managing information. Data can be input into the system manually by trained operators or transferred from other computers. The data can then be analyzed and used for several purposes, including the following:

- Checking raw materials against a work order.
- Comparing invoiced items to those that were delivered.
- Generating specifications through a bill of materials.
- Monitoring product specifications.
- Identifying accident trends and evaluating areas for correction and elimination.
- Tracking safety training.
- Charting trends in quality performance.
- Generating computerized reports to share production and quality information with workers.
- Documenting many aspects of production. Spreadsheets, databases, and reports must be produced as part of the documentation process. Spreadsheets and databases are usually built prior to production. Data is then keyed in as production moves along. Reports are usually generated later in a word processing program such as Microsoft Word™.
- Tracking and documenting substandard and scrapped parts, materials, and assemblies as required by the quality process. This is usually done through a spreadsheet in conjunction with a report generated from the data in the spreadsheet.

Once the data has been collected, it must be input into specific software programs so that it can be organized in a format that makes it useful. Many of the tracking and charting applications that computers perform are done in programs such as Microsoft Access™ and Microsoft Excel™. Access is a database program. Excel is a spreadsheet program.

Excel can be used to organize and chart simpler sets of data such as employee work schedules, daily production schedules, and daily production output. It might also be used to perform a more complex operation such as scheduling preventive maintenance based upon production hours. To produce a simple Excel spreadsheet for monitoring a production schedule and tracking progress toward a commitment, follow the steps below. See the example on p. 154.

1. Set up a column for each data type that needs to be tracked. This example uses the following: Date, Process, Time Scheduled (hours), Actual Time Required (hours), Time Difference (hours), and Net Time Loss/Gain.

2. Log the appropriate data into the corresponding columns for Date, Process, Time Scheduled (hours), and Actual Time Required (hours).

3. In the Time Difference (hours) column, set up a formula that will automatically determine and display the difference between the Time Scheduled (hours) and Actual Time Required (hours) columns for each row. This formula can easily be set up by using Excel's built-in formula tool.

4. In the Net Time Loss/Gain (hours) column (may also be set up as a row), input a formula that determines the sum of all the entries in the Time Difference (hours) column.

(continued on next page)

Date	Process	Time Scheduled (hours)	Actual Time Required (hours)	Time Difference (hours)	Net Time Loss/Gain (hours)
2/16/2005	Pre-operation safety check	0.25	−0.25	0.00	
2/16/2005	Tool setup	0.50	−0.25	0.25	
2/16/2005	Workpiece setup	0.25	−0.50	−0.25	
2/16/2005	Second safety check	0.25	−0.25	0.00	
2/16/2005	Tool power up	0.25	−0.25	0.00	
2/16/2005	Setup double-check	0.25	−0.25	0.00	
2/16/2005	1st cut 1st pass	1.25	−1.50	−0.25	
2/16/2005	1st cut 2nd pass	1.25	−1.25	0.00	
2/16/2005	1st cut 3rd pass	1.25	−1.25	0.00	
2/16/2005	1st cut 4th pass	1.25	−1.75	−0.50	
2/16/2005	1st cut 5th pass	1.75	−1.25	0.50	
2/16/2005	2nd cut 1st pass	1.50	−0.75	0.75	
2/16/2005	2nd cut 2nd pass	1.50	−1.00	0.50	
2/16/2005	2nd cut 3rd pass	2.00	−2.50	−0.50	
2/16/2005	3rd cut 1st pass	0.75	−1.00	−0.25	
2/16/2005	3rd cut 2nd pass	0.75	−1.00	−0.25	
2/16/2005	3rd cut 3rd pass	0.75	−1.00	−0.25	
2/16/2005	3rd cut 4th pass	0.75	−1.00	−0.25	
2/16/2005	3rd cut 5th pass	1.50	−2.00	−0.50	
2/16/2005	4th cut 1st pass	0.50	−0.25	0.25	
2/16/2005	Tool power down	0.25	−0.25	0.00	
2/16/2005	Workpiece takedown	0.25	−0.25	0.00	
2/16/2005	Tool breakdown	0.75	−0.50	0.25	
2/16/2005	Tool cleanup	1.25	−0.75	0.50	
2/16/2005	Work area cleanup	1.00	−1.25	−0.25	
2/16/2005	Process documentation	1.00	−2.00	−1.00	
2/16/2005	Schedule maintenance	0.75	−1.75	−1.00	
			Net Time Loss(−)/Gain(+) (hours)		−2.25

(continued on next page)

COPYRIGHT © GLENCOE/MCGRAW-HILL

Now that you have an idea of how an Excel spreadsheet is designed, try using Excel to track the daily output of a manufacturing company that produces a single product. You will need a column for each of the following:

- Date.
- Type of product.
- Number needed.
- Number produced.
- The difference between the number needed and the number produced.

Access is a software program for developing databases. It is similar to Excel in that it uses charts to organize data. While Excel may use only one or two tables, Access can be used to organize ten or more tables of related data. No single table becomes too large to be practical. The database as a whole is easily expanded and updated with new types of data. Data is not usually entered directly into the tables. Instead, it is entered into specialized forms that transfer the information into the tables. The data is usually not viewed in the tables but in reports that are more like text documents.

Access can be used in the following ways.

- To track complex data relationships.
- To track a large number of documents that must be cross-referenced.
- To organize many different types of data to analyze failure trends and identify areas that need improvement.
- To organize customer information.

Computers can also be used in conducting an audit and monitoring performance training. The following checklist presents the basic steps in collecting, reporting, and tracking the data gathered for an audit.

1. Before conducting the audit, determine the types of data to be collected.
2. Set up a spreadsheet in which to enter the data. For more complex audits involving many interrelated pieces of data, set up a database.
3. Enter the data into the spreadsheet.
4. Use a handheld calculator, the computer's calculator program, or the spreadsheet's built-in functions to summarize the data collected.
5. Use Word to build a report around the summarized data. Check with your supervisor to determine the proper format for the report.
6. Use Access to create a database for tracking and viewing documents for tool operation and qualification.

The computer can be used to maintain the documentation created for the production process. The documentation can be easily updated or archived by using the computer. Applications where it is advisable to use a computer to track and maintain records include:

- Conducting performance training.
- Documenting audits.
- Conducting quality assurance.
- Setting up the initial production run.
- Conducting general assembly procedures.
- Operating a material management system to ensure that all parts and machine capacities are available to the whole of production at all times.
- Using a time management system to schedule preventive maintenance based on production hours.
- Researching safety practices and guidelines on the Internet.
- Communicating safety information.
- Reviewing procedures programming diagnostics on computerized machine equipment.
- Obtaining customer requirements from an order tracking system to ensure quality production.

(continued on next page)

COPYRIGHT © GLENCOE/MCGRAW-HILL   MANUFACTURING APPLICATIONS   155

MEETS THE FOLLOWING (MSSC) PRODUCTION STANDARDS

P1-K2A	Raw materials are checked against work order.
P1-A1A	Use computerized manufacturing system for Bill of Materials specifications and general assembly procedures.
P1-A1B	Use computer to input quality data collected.
P1-A1C	Obtain customer requirements from order tracking system.
P1-A1D	Use a material management system in order to ensure all parts and machine capacity is available to all production.
P1-A1E	Use PC to document initial setup of production run.
P2-A1A	Use Time Management System to schedule preventive maintenance based on production hours.
P2-A15F	Review procedures for programming diagnostics on computer drive machine equipment.
P3-A1B	Use computerized data collection to identify accident trends/areas that need to be evaluated for correction and elimination.
P3-A1E	Use computer to track safety training.
P4-A1C	Use calculator to conduct audits and ensure product quality at different stages of the production cycle.
P4-A1D	Use PC to create and maintain audit documentation.
P4-A1E	Use Access data base to collect field and in-house data in order to identify failure trends and point to areas for improvement opportunities.
P5-A1C	Post production schedule on Excel spreadsheet to monitor and track progress to commitment.
P5-A1E	Use computerized reports to share production and quality information with production workers.
P5-A16C	Compare invoice to delivery.
P6-A1B	Track daily output on spreadsheet.
P7-A1C	Use computer to document and monitor performance training.
P7-A1D	Use technology to research safety practices and guidelines.
P7-A1E	Use computer and other telecomm equipment to communicate safety information.
P7-A1F	Use document database to view documents for tool operation and qualification.
P8-O3A	Skill in using computing systems to document and track substandard and scrapped parts, materials, and assemblies as required by quality processes.
P8-A1B	Use PC to trend quality performance.

COPYRIGHT © GLENCOE/McGRAW-HILL

Checking Machine Guards

There are many possible machinery-related injuries: crushed hands, severed limbs, and blindness, for example. Safeguards are essential for protecting workers from needless and preventable injuries. Any machine part, function, or process that can cause injury must be safeguarded. When the operation of a machine or accidental contact with it can injure the operator or others in the vicinity, the hazards must be either controlled or eliminated.

Power tools and production equipment include safety systems designed to prevent worker contact with moving or cutting parts. These safety systems protect workers by physically preventing contact or by preventing the tool from operating when safety procedures are not followed. Most of the safety systems are regulated and enforced by the Occupational Safety and Health Administration (OSHA).

Protective Machine Guards

Protective machine guards are designed to prevent accidents by providing a barrier between the worker and moving parts of the equipment. The following are examples of machine guards.

- Large power tools with moving parts that create cutting, pinching, or crushing hazards usually have a plastic or metal shield or some other mechanism to prevent contact with the moving part.
- Some guards are designed to prevent entanglement with moving parts such as shafts and gears.
- Some guards prevent the operator from being struck by material ejected from the machine.

There are a number of different types of guards designed for different purposes. A safety device that incorporates a guard may perform one of several functions. It may:

- Stop the machine if a hand or any part of the body enters the danger area.
- Restrain or move the operator from the danger area during operation.
- Require the operator to use both hands on machine controls, forcing hands and body out of the danger area.

- Provide a barrier that is synchronized with the operating cycle of the machine to prevent entry to the danger area during the hazardous part of the cycle.

Fixed guards. A fixed guard, which has no moving parts, forms a barrier to prevent contact between moving machinery and the body. The guard is a permanent part of the machine. It may be made of sheet metal, screen, wire cloth, or plastic. Examples of fixed guards are covers over moving gears and metal plates that prevent hair or clothing from getting between rollers. Railings and other physical barriers are also fixed guards. In addition to protecting workers from moving parts, fixed guards prevent contact with electrical charges, chemical hazards, and heat hazards.

Automatic guards. These guards are designed to move the operator away from the hazard point before injury can occur during part of the operation. They are sometimes called sweep guards because the guard sweeps the operator away from the tool. Automatic guards are often used in power presses. The guard gets its motion from a linkage to the operating part, and it multiplies the movement of the part. A moving barrier forces the person or body part away from the contact point. Automatic guards are generally not used in equipment that moves very rapidly, because the high speed of the guard itself would be a potential hazard. To be effective, automatic guards must be properly adjusted to keep sufficient distance between the operator and the moving part. All the parts of the guard must move at the correct time.

(continued on next page)

Safety First

Using the guard is necessary for safe operation. Guards should never be removed or held away from the moving part during operation. Before using equipment that has any type of guard, inspect the guard. Make sure that it has not been damaged or moved out of place and that no parts of the guard are loose.

Safety Inspection. Safety systems will protect workers only if they are operating properly. *Each time* you use a tool or piece of equipment that includes a safety system, inspect the system to be sure that it is not damaged or out of adjustment before you begin working.

Maintenance. If you have been trained in maintenance of the safety system, you may be able to repair or adjust it. If you have not been trained, do not attempt to maintain the system. Instead, notify your supervisor or maintenance personnel that the guard or alarm system is out of order. You should not use a piece of equipment if the guards or other protective devices are not in place and operating properly.

Training. Before using this equipment, operators should attend safety or OSHA training. If you are training new operators to use the equipment, explain the location and function of the safety devices.

Safety Audits. Equipment and systems must also be audited on a regular basis to ensure that they are working properly. This would include making sure that there are no bypasses of safety guards.

Safety Audit Checklist for Machines

The following checklist has been adapted from OSHA information. It can be used as part of a safety audit to determine whether machine guards are being used effectively.

Requirements for All Safeguards

❑ The safeguards provided meet the minimum OSHA requirements.

❑ The safeguards prevent workers' hands, arms, and other body parts from making contact with dangerous moving parts.

❑ The safeguards are firmly secured and not easily removable or tampered with.

❑ The safeguards are made of durable material that will withstand the conditions of normal use.

❑ The safeguards ensure that no object will fall into the moving parts, creating a projectile.

❑ The safeguard creates no new hazard. There are no jagged edges, shear points, or unfinished surfaces that could cause laceration. Edges of guards are rolled or bolted to eliminate sharp edges.

❑ The safeguards create no interference. They permit safe, comfortable, and relatively easy operation of the machine and relieve workers' concerns about injury.

❑ The machine can be oiled without removing the safeguard.

❑ There is a system for shutting down the machinery before safeguards are removed.

Mechanical Hazards

Point of Operation. This is the point where work is performed on the material, such as cutting, shaping, boring, or forming of stock.

❑ A point-of-operation safeguard is provided for the machine.

❑ The safeguard keeps the operator's hands, fingers, and body out of the danger area.

❑ There is no evidence that the safeguards have been tampered with or removed.

(continued on next page)

COPYRIGHT © GLENCOE/MCGRAW-HILL

Power Transmission Apparatus. This includes all components of the mechanical system that transmit energy to the part of the machine performing the work. These components include flywheels, pulleys, belts, connecting rods, couplings, cams, spindles, chains, cranks, and gears.

❏ There are no unguarded gears, sprockets, pulleys, or flywheels on the apparatus.

❏ There are no exposed belts or chain drives.

❏ There are no exposed setscrews, keyways, or collars.

❏ Starting and stopping controls are within easy reach of the operator.

❏ There are separate controls provided for more than one operator.

Other Moving Parts. These are all parts of the machine that move while the machine is working. They can include reciprocating, rotating, and transverse moving parts, as well as feed mechanisms and auxiliary parts of the machine.

❏ Safeguards are provided for all hazardous moving parts of the machine, including auxiliary parts.

Nonmechanical Hazards

❏ Appropriate measures have been taken to safeguard workers against noise hazards.

❏ Special guards, enclosures, ventilation, and personal protective equipment (PPE) have been provided, where necessary, to protect workers from exposure to harmful substances.

Electrical Hazards

❏ The machine is installed in accordance with National Fire Protection Association and National Electrical Code requirements.

❏ There are no loose conduit fittings.

❏ The machine is properly grounded.

❏ The power supply is correctly fused and protected.

❏ Workers do not receive minor shocks while operating the machine.

Training

❏ Operators and maintenance workers have been trained about the hazards associated with particular machines.

❏ Operators and maintenance workers have the necessary training in how to use the safeguards.

❏ Operators and maintenance workers have been trained in where the safeguards are located, how they provide protection, and what hazards they protect against.

❏ Operators and maintenance workers have been trained in how and under what circumstances guards can be removed.

❏ Workers have been trained in the procedures to follow if they notice guards that are damaged, missing, or inadequate.

Personal Protective Equipment & Proper Clothing

❏ The operator is required to wear PPE.

❏ Required PPE is appropriate for the job, in good condition, clean and sanitary, and stored carefully when not in use.

❏ The operator is dressed safely for the job (i.e., no loose-fitting clothing or jewelry).

(continued on next page)

Machinery Maintenance & Repair

❏ Maintenance workers have received up-to-date instruction on the machines they service.

❏ Maintenance workers lock out the machine from its power sources before beginning repairs.

❏ Where several maintenance persons work on the same machine, multiple lockout devices are used.

❏ Maintenance workers use appropriate and safe equipment in their repair work.

❏ The maintenance equipment itself is properly guarded.

❏ Maintenance and servicing workers are trained in the requirements of lockout/tagout hazard, and the procedures for lockout/tagout exist before they attempt their tasks.

MEETS THE FOLLOWING (MSSC) PRODUCTION STANDARDS

P3-O2E	Knowledge of clothing and PPE that should be worn to ensure safety.
P3-A3B	Select proper PPE for the job to prevent injuries.
P7-K3E	Equipment is audited to ensure there are no bypasses of safety guards.
P7-O1A	Knowledge of government policies, procedures, and regulations governing the safe use of equipment.
P7-04B	Knowledge of equipment safety systems to verify they are operating properly.
P7-04D	Knowledge of PPE that should be worn.
P7-04F	Skill in evaluating workplace safety using safety audit processes.
P7-A11E	Attend safety and OSHA training on equipment.
P7-A12C	Explain to new operators the location and function of safety devices on the equipment they are using.

COPYRIGHT © GLENCOE/McGRAW-HILL

Checking Interlocks & Emergency Stop Controls

There are many possible machinery-related injuries: crushed hands, severed limbs, and blindness, for example. Safeguards are essential for protecting workers from needless and preventable injuries. Any machine part, function, or process that can cause injury must be safeguarded. When the operation of a machine or accidental contact with it can injure the operator or others in the vicinity, the hazards must be either controlled or eliminated.

Power tools and production equipment include safety systems designed to prevent contact with moving or cutting parts. These safety systems protect workers by physically preventing contact or by preventing the tool from operating when safety procedures are not followed. Most of the safety systems are regulated and enforced by the Occupational Safety and Health Administration (OSHA).

Some tools and equipment have warning systems to alert you to danger. Some systems actually shut down a piece of equipment if a hazardous situation is detected.

Interlocks

An interlock is a guard system that prevents a tool from being used if certain conditions have not been met. For example, a guard may have to be in place, two switches may have to be depressed at the same time, or an area may have to be cleared of people before a device can be operated. Interlocks include simultaneous control devices, safety mats, and light curtains.

Simultaneous Control Devices. These devices allow the equipment to operate only if the operator's hands are in a position where they are not in danger. This type of system is frequently used in large presses and in cutting machines. The operator must press two buttons simultaneously before the machine will start. The buttons are placed in such a way that both hands are required. This ensures that the machine cannot be operated if the operator's hands are within a crushing or cutting area.

Safety Mats. These pressure-sensing floor coverings will not allow the machine to operate if a weight is on top of the mat. The advantage of safety mats is that one operator cannot start a machine while another person is in the hazard area.

Light Curtains. These devices use photoelectronic sensors to sense obstructions in a pathway. If someone reaches across the beam of light into a monitored area, the safety system will shut off the equipment. The machine cannot be restarted until the area is clear of obstructions. Light curtains are frequently used for robotic systems. Once the system is shut off, it cannot restart until someone manually pushes a restart button after clearing the area.

- If equipment that includes an interlock operates without the interlock being activated, the equipment should not be used.
- The failure of the interlock must be reported and corrected before the machine is operated.

Emergency Stop Controls

An emergency stop control is an electrical or mechanical control used to stop a machine when an emergency occurs. An emergency stop control overrides all other machine controls. It is usually placed in a very visible position as a large red button. Once an emergency stop control has been activated, the equipment must be restarted using a separate control.

Safety trip controls provide a quick way to stop the machine in an emergency situation. For example, a pressure-sensitive body bar can be placed

(continued on next page)

in front of the moving parts. If the operator trips, loses balance, or is drawn toward the machine, applying pressure to the bar will stop the operation. The positioning of the bar is critical. It must stop the machine before a part of the worker's body reaches the danger area. Trip controls must be manually reset to restart the machine. Just releasing the control to restart the machine will not ensure that the worker is out of danger when the machine restarts.

Safety First

Safety Inspection. Safety systems will protect workers only if they are operating properly. *Each time* you use a tool or piece of equipment that includes a safety system, inspect the system to be sure that it is not damaged or out of adjustment before you begin working.

Maintenance. If you have been trained in maintenance of the safety system, you may be able to repair or adjust it. If you have not been trained, do not attempt to maintain the system. Instead, notify your supervisor or maintenance personnel that the guard or alarm system is out of order. You should not use a piece of equipment if the guards or other protective devices are not in place and operating properly.

Training. Before using this equipment, operators should attend safety or OSHA training. If you are training new operators to use the equipment, explain the location and function of the safety devices.

Safety Audits. Equipment and systems must also be audited on a regular basis to ensure that they are working properly. This would include making sure that there are no bypasses of safety guards.

Safety Audit Checklist for Machines

The following checklist has been adapted from OSHA information. It can be used as part of a safety audit to determine whether machine guards are being used effectively.

Requirements for All Safeguards

❑ The safeguards provided meet the minimum OSHA requirements.

❑ The safeguards prevent workers' hands, arms, and other body parts from making contact with dangerous moving parts.

❑ The safeguards are firmly secured and not easily removable or tampered with.

❑ The safeguards are made of durable material that will withstand the conditions of normal use.

❑ The safeguards ensure that no object will fall into the moving parts, creating a projectile.

❑ The safeguard creates no new hazard. There are no jagged edges, shear points, or unfinished surfaces that could cause laceration. Edges of guards are rolled or bolted to eliminate sharp edges.

❑ The safeguards create no interference. They permit safe, comfortable, and relatively easy operation of the machine and relieve workers' concerns about injury.

❑ The machine can be oiled without removing the safeguard.

❑ There is a system for shutting down the machinery before safeguards are removed.

Mechanical Hazards

Point of Operation. This is the point where work is performed on the material, such as cutting, shaping, boring, or forming of stock.

❑ A point-of-operation safeguard is provided for the machine.

❑ The safeguard keeps the operator's hands, fingers, and body out of the danger area.

❑ There is no evidence that the safeguards have been tampered with or removed.

(continued on next page)

COPYRIGHT © GLENCOE/McGRAW-HILL

Power Transmission Apparatus. This includes all components of the mechanical system that transmit energy to the part of the machine performing the work. These components include flywheels, pulleys, belts, connecting rods, couplings, cams, spindles, chains, cranks, and gears.

❑ There are no unguarded gears, sprockets, pulleys, or flywheels on the apparatus.

❑ There are no exposed belts or chain drives.

❑ There are no exposed setscrews, keyways, or collars.

❑ Starting and stopping controls are within easy reach of the operator.

❑ There are separate controls provided for more than one operator.

Other Moving Parts. These are all parts of the machine that move while the machine is working. They can include reciprocating, rotating, and transverse moving parts, as well as feed mechanisms and auxiliary parts of the machine.

❑ Safeguards are provided for all hazardous moving parts of the machine, including auxiliary parts.

Nonmechanical Hazards

❑ Appropriate measures have been taken to safeguard workers against noise hazards.

❑ Special guards, enclosures, ventilation, and personal protective equipment (PPE) have been provided, where necessary, to protect workers from exposure to harmful substances.

Electrical Hazards

❑ The machine is installed in accordance with National Fire Protection Association and National Electrical Code requirements.

❑ There are no loose conduit fittings.

❑ The machine is properly grounded.

❑ The power supply is correctly fused and protected.

❑ Workers do not receive minor shocks while operating the machine.

Training

❑ Operators and maintenance workers have been trained about the hazards associated with particular machines.

❑ Operators and maintenance workers have the necessary training in how to use the safeguards.

❑ Operators and maintenance workers have been trained in where the safeguards are located, how they provide protection, and what hazards they protect against.

❑ Operators and maintenance workers have been trained in how and under what circumstances guards can be removed.

❑ Workers have been trained in the procedures to follow if they notice guards that are damaged, missing, or inadequate.

Personal Protective Equipment & Proper Clothing

❑ The operator is required to wear PPE.

❑ Required PPE is appropriate for the job, in good condition, clean and sanitary, and stored carefully when not in use.

❑ The operator is dressed safely for the job (i.e., no loose-fitting clothing or jewelry).

(continued on next page)

Machinery Maintenance & Repair

☐ Maintenance workers have received up-to-date instruction on the machines they service.

☐ Maintenance workers lock out the machine from its power sources before beginning repairs.

☐ Where several maintenance persons work on the same machine, multiple lockout devices are used.

☐ Maintenance workers use appropriate and safe equipment in their repair work.

☐ The maintenance equipment itself is properly guarded.

☐ Maintenance and servicing workers are trained in the requirements of lockout/tagout hazard, and the procedures for lockout/tagout exist before they attempt their tasks.

MEETS THE FOLLOWING (MSSC) PRODUCTION STANDARDS

P1-K4C	Operations are performed safely.
P3-O2E	Knowledge of clothing and PPE that should be worn to ensure safety.
P3-A3B	Select proper PPE for the job to prevent injuries.
P7-K3E	Equipment is audited to ensure there are no bypasses of safety guards.
P7-O1A	Knowledge of government policies, procedures, and regulations governing the safe use of equipment.
P7-O4B	Knowledge of equipment safety systems to verify they are operating properly.
P7-O4D	Knowledge of PPE that should be worn.
P7-O4F	Skill in evaluating workplace safety using safety audit processes.
P7-A11E	Attend safety and OSHA training on equipment.
P7-A12C	Explain to new operators the location and function of safety devices on the equipment they are using.

COPYRIGHT © GLENCOE/McGRAW-HILL

Checking Alarms, Safety Valves & Rupture Disks

There are many possible machinery-related injuries: crushed hands, severed limbs, and blindness, for example. Safeguards are essential for protecting workers from needless and preventable injuries. Any machine part, function, or process that can cause injury must be safeguarded. When the operation of a machine or accidental contact with it can injure the operator or others in the vicinity, the hazards must be either controlled or eliminated.

Power tools and production equipment include safety systems designed to prevent contact with moving and cutting parts. These safety systems protect workers by physically preventing contact or by preventing the tool from operating when safety procedures are not followed. Most of the safety systems are regulated and enforced by the Occupational Safety and Health Administration (OSHA).

Some tools and equipment have warning systems to alert you to danger. These systems include visual or audible alarms to warn of immediate danger. Some release pressure if a hazardous situation is detected.

Alarms

Many safety systems include visual or auditory alarms to alert workers of danger. The following are examples of alarm systems.

- A photoelectronic device such as a light curtain might cause a bell to ring if an obstruction is detected prior to shutting down the equipment.

- Alarms can also warn of hazardous conditions such as dangerous concentrations of chemicals. A carbon dioxide detector may cause a light to flash and a siren to sound if the concentration of carbon dioxide in the air reaches a dangerous level.

- Other alarm systems warn of potential danger. The beeping sound that you hear when someone backs up a forklift or other device in which the operator may have a limited field of vision warns you to get out of the way.

- When a large piece of equipment such as a conveyor belt is turned on, an alarm may sound for several seconds before motion begins. This warns other workers to clear the area and avoid the hazard.

Anyone working in an area where an alarm may be activated must be trained to understand and react appropriately to the alarm.

Safety Valves & Rupture Disks

Manufacturing plants in industries that work with materials under high pressure—such as those in the chemical, plastics and rubber, petroleum products, and food and beverage industries—use safety devices to avoid excess pressure. These devices include safety valves and rupture disks.

Safety Valve. A safety valve opens if the pressure exceeds the design limit of the equipment.

Rupture Disk. A rupture disk is a device that breaks if a sudden pressure rise causes pressure to exceed the safe limit. When a rupture disk breaks, the material under pressure may be directed to a vessel that provides containment.

- These devices must be inspected on a regular schedule following procedures provided by the manufacturer.

- Before using a piece of equipment that incorporates safety valves or rupture disks, check the inspection tag to be certain that the inspection is current.

(continued on next page)

Safety First

Safety Inspection. Safety systems will protect workers only if they are operating properly. *Each time* you use a tool or piece of equipment that includes a safety system, inspect the system to be sure that it is not damaged or out of adjustment before you begin working.

Maintenance. If you have been trained in maintenance of the safety system, you may be able to repair or adjust it. If you have not been trained, do not attempt to maintain the system. Instead, notify your supervisor or maintenance personnel that the guard or alarm system is out of order. You should not use a piece of equipment if the guards or other protective devices are not in place and operating properly.

Training. Before using this equipment, operators should attend safety or OSHA training. If you are training new operators to use the equipment, explain the location and function of the safety devices.

Safety Audits. Equipment and systems must also be audited on a regular basis to ensure that they are working properly. This would include making sure that there are no bypasses of safety guards.

Safety Audit Checklist for Machines

The following checklist has been adapted from OSHA information. It can be used as part of a safety audit to determine whether machine guards are being used effectively.

Requirements for All Safeguards

❑ The safeguards provided meet the minimum OSHA requirements.

❑ The safeguards prevent workers' hands, arms, and other body parts from making contact with dangerous moving parts.

❑ The safeguards are firmly secured and not easily removable or tampered with.

❑ The safeguards are made of durable material that will withstand the conditions of normal use.

❑ The safeguards ensure that no object will fall into the moving parts, creating a projectile.

❑ The safeguard creates no new hazard. There are no jagged edges, shear points, or unfinished surfaces that could cause laceration. Edges of guards are rolled or bolted to eliminate sharp edges.

❑ The safeguards create no interference. They permit safe, comfortable, and relatively easy operation of the machine and relieve workers' concerns about injury.

❑ The machine can be oiled without removing the safeguard.

❑ There is a system for shutting down the machinery before safeguards are removed.

Mechanical Hazards

Point of Operation. This is the point where work is performed on the material, such as cutting, shaping, boring, or forming of stock.

❑ A point-of-operation safeguard is provided for the machine.

❑ The safeguard keeps the operator's hands, fingers, and body out of the danger area.

❑ There is no evidence that the safeguards have been tampered with or removed.

(continued on next page)

   COPYRIGHT © GLENCOE/McGRAW-HILL

Power Transmission Apparatus. This includes all components of the mechanical system that transmit energy to the part of the machine performing the work. These components include flywheels, pulleys, belts, connecting rods, couplings, cams, spindles, chains, cranks, and gears.

❑ There are no unguarded gears, sprockets, pulleys, or flywheels on the apparatus.

❑ There are no exposed belts or chain drives.

❑ There are no exposed setscrews, keyways, or collars.

❑ Starting and stopping controls are within easy reach of the operator.

❑ There are separate controls provided for more than one operator.

Other Moving Parts. These are all parts of the machine that move while the machine is working. They can include reciprocating, rotating, and transverse moving parts, as well as feed mechanisms and auxiliary parts of the machine.

❑ Safeguards are provided for all hazardous moving parts of the machine, including auxiliary parts.

Nonmechanical Hazards

❑ Appropriate measures have been taken to safeguard workers against noise hazards.

❑ Special guards, enclosures, ventilation, and personal protective equipment (PPE) have been provided, where necessary, to protect workers from exposure to harmful substances.

Electrical Hazards

❑ The machine is installed in accordance with National Fire Protection Association and National Electrical Code requirements.

❑ There are no loose conduit fittings.

❑ The machine is properly grounded.

❑ The power supply is correctly fused and protected.

❑ Workers do not receive minor shocks while operating the machine.

Training

❑ Operators and maintenance workers have been trained about the hazards associated with particular machines.

❑ Operators and maintenance workers have the necessary training in how to use the safeguards.

❑ Operators and maintenance workers have been trained in where the safeguards are located, how they provide protection, and what hazards they protect against.

❑ Operators and maintenance workers have been trained in how and under what circumstances guards can be removed.

❑ Workers have been trained in the procedures to follow if they notice guards that are damaged, missing, or inadequate.

Personal Protective Equipment & Proper Clothing

❑ The operator is required to wear PPE.

❑ Required PPE is appropriate for the job, in good condition, clean and sanitary, and stored carefully when not in use.

❑ The operator is dressed safely for the job (i.e., no loose-fitting clothing or jewelry).

(continued on next page)

Machinery Maintenance & Repair

❑ Maintenance workers have received up-to-date instruction on the machines they service.

❑ Maintenance workers lock out the machine from its power sources before beginning repairs.

❑ Where several maintenance persons work on the same machine, multiple lockout devices are used.

❑ Maintenance workers use appropriate and safe equipment in their repair work.

❑ The maintenance equipment itself is properly guarded.

❑ Maintenance and servicing workers are trained in the requirements of lockout/tagout hazard, and the procedures for lockout/tagout exist before they attempt their tasks.

MEETS THE FOLLOWING PRODUCTION STANDARDS

P1-K4C	Operations are performed safely.
P3-O2E	Knowledge of clothing and PPE that should be worn to ensure safety.
P3-A3B	Select proper PPE for the job to prevent injuries.
P7-K3E	Equipment is audited to ensure there are no bypasses of safety guards.
P7-O1A	Knowledge of government policies, procedures, and regulations governing the safe use of equipment.
P7-O4B	Knowledge of equipment safety systems to verify they are operating properly.
P7-O4D	Knowledge of PPE that should be worn.
P7-O4F	Skill in evaluating workplace safety using safety audit processes.
P7-A11E	Attend safety and OSHA training on equipment.
P7-A12C	Explain to new operators the location and function of safety devices on the equipment they are using.

COPYRIGHT © GLENCOE/MCGRAW-HILL

Monitoring Systems for Leaks

Facilities that handle hazardous liquid or gaseous materials are required to have a leak monitoring program to prevent health and environmental hazards. This ensures that no toxic or hazardous material is escaping from a system or a piece of equipment. Even when handling materials that are not classified as hazardous, leaks can create other hazards, such as slippery floors. There are several methods for detecting leaks.

Electronic Leak Detectors. Gas leaks are monitored using an electronic leak detector with a sensor designed and calibrated to detect the type of material in the system. Electronic detectors should be used only by workers trained in their use. Follow the standard operating procedure (SOP) or the manual provided by the manufacturer of the leak detector. Electronic leak detectors may also be used for liquid leaks, particularly for liquids with a low vapor pressure.

Soap Solutions. Gaseous leaks that do not involve hazardous materials, particularly leaks of air, can be monitored using a soap solution. The solution is applied to lightly coat the fitting and observed to determine whether bubbles form. This method, however, is generally effective only for low-volume leaks. It is not suitable for checking hazardous materials or systems which could have an internal vacuum.

Observation. Leaks of liquid chemicals can often be detected by observation of fittings, valves, and sealed joints for escaping liquid.

Safety Precautions

❑ When checking for leaks of materials that are hazardous, under pressure, or at high temperature, wear appropriate personal protective equipment.

❑ If you detect unsafe conditions, report or correct the hazard immediately.

❑ Whenever possible, approach equipment containing hazardous materials from upwind.

❑ Equipment in a sheltered area should be approached slowly from a more open area.

❑ If using an electronic detector, use care when monitoring sources, such as valves and pumps, that handle heavy liquid streams at high temperatures. Materials can condense in the probe and detector, affecting the instrument response.

❑ A rigid probe tip should not be used near a rotating pump shaft because it could break if it touches the moving shaft. A flexible tip is usually added to the end of the rigid probe when sampling pumps.

Checking a System for Leaks

Leaks tend to occur at joints between components of a system, at valves, and around pump seals.

1. Proceed from one end of a flow route to the opposite end, checking each possible leak site.

2. If there are numerous routes, mark off segments on a chart or use a tag to indicate which parts of the system have been checked.

3. Use appropriate tags to mark any leaks if you are not repairing them during the checking procedure.

4. Leaks from valves occur primarily from the valve stem packing gland. The normal procedure is to examine all around the valve stem. If using an electronic leak detector, hold the probe at the stem-packing interface.

5. Emissions from pumps occur from the pump shaft seal used to isolate the fluid from the atmosphere. If the protective housing prevents examining the entire rotating shaft, sample all accessible portions.

(continued on next page)

6. Other points on the pump or compressor housing where leaks could occur include flanges and other connections. They typically occur at the flange or connector sealing interface. When using a detector, the probe should be positioned at the outer edge of the flange-gasket interface. Check the entire circumference of the flange.

7. Usually, the configuration of pressure-relief devices prevents sampling at the sealing interface. If the pressure-relief device has an enclosed discharge pipe or extension, a sample probe should be placed at the center of the exhaust area. Also, sample at any "weep hole" at the bottom of such devices.

8. The probe tip should be routinely checked and cleaned out as necessary. Instrument background readings and return-to-zero rates should be monitored periodically.

9. Mark any open-ended lines (lines missing a plug or an end cap) for repair, even if no leak is detected at the time of monitoring.

10. Document all leaks and equipment needing repair using appropriate forms.

11. File records according to the SOP.

MEETS THE FOLLOWING (MSSC) PRODUCTION STANDARD

P7-04E	Skill in performing leak checks to determine if toxic or hazardous material is escaping from a piece of equipment.

COPYRIGHT © GLENCOE/MCGRAW-HILL

Following Lockout/Tagout Procedures

Lockout/tagout is a series of procedures designed to protect workers from the release of hazardous, or stored, energy during maintenance and other shutdown procedures. This energy can be electrical, mechanical, chemical, or thermal. For example, a high-voltage conduit that could arc to a nearby person or a spring-loaded device under tension could potentially deliver hazardous energy to a worker.

Shutdowns are sometimes planned well in advance. In other cases, workers may have short notice. Sometimes a shutdown may need to proceed immediately because of an emergency. In all cases, workers must follow standard operating procedures (SOPs) to safely turn off the equipment.

Lockout. The SOPs for locking out a machine or other device involve completely disconnecting the device from all energy sources and then applying a substantial locking mechanism that prevents the device from being reconnected to any energy source.

Tagout. The SOPs for tagging out a device involve applying a special tag to warn others not to reconnect the device to an energy source.

The two procedures are generally used in common to maximize the safety of the worker who is servicing a piece of equipment. The SOPs ensure that:

- The process proceeds safely.
- The process avoids damage to the equipment.

For example, in the chemical, petroleum, and food and beverage subindustries, contents of a vessel may be very hot. If the stirrer, or agitator, is turned off prior to cooling the contents, material can begin to boil in an uncontrolled way. This causes a risk of leaks or spills of hot material, a possible cause of burn injuries. Mechanical energy stored in springs or ratchets poses the risk of crushing or cutting injuries.

Depending on the plant design, there may be one generalized procedure for all of the equipment or individual procedures for different types of equipment. In either case, there are several steps that must be performed, in the correct order. The primary goal of these steps is to remove any active or stored energy from the system.

Lockout/Tagout Procedures

The basic steps required by OSHA for performing lockout/tagout on equipment are:

1. Notify affected workers that lockout/tagout is about to occur on a specific piece of machinery or equipment. Prepare for shutdown by reviewing details of the energy source, hazards, and specific control procedures.

2. Shut down the machine or equipment using normal stopping or rundown procedures for that machine.

3. Isolate the equipment from the energy source or sources. These may include electricity, hydraulic pressure, pressurized steam, residual mechanical energy, compressed gas lines, charged chemical lines, and chemical drain lines. Isolating the equipment from its energy source may involve turning off such items as the operating control, a line valve, or an electrical circuit breaker.

4. Apply the lockout/tagout devices to the energy-isolating devices. For example, a padlock can be placed through holes so that switch handles are locked in the "off" position and cannot be moved. These devices must be substantial and durable.

(continued on next page)

5. Safely release any potentially hazardous stored or residual energy. The specific procedures to be followed depend on the type of energy and how it is stored. Examples of energy release include returning springs to a normal position, bleeding down hydraulic systems, and cooling heated parts or chemicals. The machine must be in a zero-energy state. If there is any chance that stored energy may reaccumulate, verification of isolation must be continued for the duration of the shutdown.

6. Verify that energy control measures are effective before doing any work on the equipment or leaving it unattended. For example, turn switches or start buttons to the "on" position to ensure that the power is actually isolated. Then return them to the "off" position.

When work is completed, take the following steps:

1. Verify that the equipment is intact and working properly.

2. Keep all workers at a safe distance from the equipment.

3. Ensure that the lockout/tagout devices are removed from each energy-isolating device by the employee who applied the device in the first place.

Lockout/Tagout Policies

- A lockout/tagout program specific to the workplace should be developed, written, and distributed to the workforce.

- Each worker should receive written and verbal confirmation on the lockout/tagout procedures.

- Workers should understand their responsibilities within the lockout/tagout program.

- Training should be conducted to ensure that the workforce is familiar with lockout/tagout procedures.

- Lockout/tagout procedures that are part of company or federal safety policy should be followed at all times. Failure to follow these procedures can result in injury or death.

MEETS THE FOLLOWING (MSSC) PRODUCTION STANDARDS

P2-K1H	All safety procedures are followed when doing repairs.
P2-O4D	Knowledge of lockout/tagout policies and procedures.
P3-O1H	Knowledge of lockout/tagout requirements.
P7-K3F	All regulatory and company safety procedures are followed including lockout/tagout, confined spaces, and ergonomics.
P7-O3E	Skill in training other workers in proper safety procedures during maintenance process.
P7-O4C	Knowledge of how to prevent unsafe shutdown of equipment.
P7-A7A	Shut down out-of-compliance equipment to assure unsafe practices or out-of-compliance does not occur.
P7-A14E	Develop a written lockout/tagout program so everyone knows what is going to take place.

 COPYRIGHT © GLENCOE/McGRAW-HILL

Checking Confined Spaces

Confined spaces are an issue in many areas of the manufacturing industry. Confined spaces are "confined" in the sense that they substantially limit the activities of the workers who must enter, work in, and exit them. The Occupational Safety and Health Administration (OSHA) defines a confined space as "a compartment of small size and limited access such as a double bottom tank, cofferdam, or other space which by its small size and confined nature can readily create or aggravate a hazardous exposure." Given their potential to cause injury, special rules were created to deal with the problem of confined spaces. A confined space should always be checked and verified as safe before being entered. The confined space pre-entry checklist that follows presents a set of precautions that will help verify the safety of the confined space. This is only a general checklist.

☐ Are confined spaces thoroughly emptied of all corrosive and hazardous substances, such as acids or caustics, before being entered?

☐ Are all lines to a confined space that contain inert, toxic, flammable, or corrosive materials valved off and blanked and disconnected and separated before entry?

☐ Are all impellers, agitators, and other moving parts and equipment inside confined spaces locked-out if they present a hazard?

☐ Is natural or mechanical ventilation provided prior to confined space entry?

☐ Are appropriate atmospheric tests performed to check for oxygen deficiency, toxic substances, and explosive concentrations in the confined space before entry?

☐ Is adequate illumination provided for the work to be performed in the confined space?

☐ Is the atmosphere inside the confined space frequently tested or continuously monitored during conduct of work?

☐ Is there an assigned safety standby employee outside the confined space whose sole responsibility is to watch the work in progress, sound an alarm if necessary, and render assistance?

☐ Is the standby employee provided with the appropriate personal protective equipment and trained to handle an emergency?

☐ Are the standby employee and other employees prohibited from entering the confined space without lifelines and respiratory equipment if there is any question as to the cause of an emergency?

☐ Is approved respiratory equipment required if the atmosphere inside the confined space cannot be made acceptable?

☐ Is all portable electrical equipment used inside confined spaces either grounded and insulated or equipped with ground fault protection?

☐ Before gas welding or burning is started in a confined space, are hoses checked for leaks, compressed gas bottles forbidden inside the confined space, and torches lit only outside of the confined space?

☐ Is the confined space tested for an explosive atmosphere each time before a lit torch is taken into the confined space?

☐ If employees will be using oxygen-consuming equipment—such as salamanders, torches, and furnaces—in a confined space, is sufficient air provided to assure combustion without reducing the oxygen concentration of the atmosphere below 19.5% by volume?

☐ Whenever combustion-type equipment is used in a confined space, are provisions made to ensure that the exhaust gases are vented outside the enclosure?

(continued on next page)

❏ Is each confined space checked for decaying vegetation or animal matter that may produce methane?

❏ Is the confined space checked for industrial waste that could contain toxic properties?

❏ If the confined space is below ground and near areas where motor vehicles will be operating, is it possible for vehicle exhaust or carbon monoxide to enter the space?

Each company or industry may have its own safety requirements and policies regarding confined spaces. These requirements and policies should be readily available to workers who may have to work in a confined space in the course of their employment. Be sure to review all safety procedures relating to working in confined spaces. Be sure you understand these procedures. Be sure to follow all safety procedures when working in confined spaces. Failure to do so may result in injury or death.

MEETS THE FOLLOWING PRODUCTION STANDARD

P7-K3F	All regulatory and company safety procedures are followed including lockout & tagout, confined space and ergonomics.

COPYRIGHT © GLENCOE/McGRAW-HILL

Performing Maintenance on a Drill Press

The following general checklist outlines the basic procedures and schedules for maintenance of a drill press. This checklist is a guide and is by no means comprehensive. Check the owner's manual of your particular machine for a more detailed set of maintenance procedures. The appropriate individual should provide training on the procedures needed to maintain equipment.

Daily Maintenance

❑ Visually check all electrical interfaces such as switches and wiring. Arrange for the immediate repair of problems such as damaged switches and cracks in wire sheathing.

❑ Check that the appropriate personal protective equipment (PPE) is available and in good working order.

❑ Check that all safety guards are in place and functioning properly.

❑ Make sure the workstation is clean and free of all obstacles and obstructions.

❑ Check the manufacturer's specifications and lubricate the machine accordingly.

Weekly Maintenance

❑ Visually check all electrical interfaces such as switches and wiring. Arrange for the immediate repair of problems such as damaged switches and cracks in wire sheathing.

❑ Check the chuck key to ensure that it works correctly.

❑ Check the column safety collar to ensure that it is in good working order.

❑ Check that the table adjustment moves freely.

❑ Check that drill bits are available and in good working order.

❑ Lubricate the column and the rest of the machine according to the manufacturer's directions.

Quarterly Maintenance

❑ Make sure spare parts such as belts, drills, chucks, and cutters are available.

❑ Following manufacturer's specifications, lubricate all moving parts.

❑ Protect bare metal surfaces with a petroleum-based lubricant/solvent.

Six-Month Maintenance

❑ Clean waste from housings, cabinets, and other restricted access spaces.

❑ Check belts for flaws.

❑ Make sure that belts are set at the proper tension.

❑ Check pulley alignment.

❑ Examine all mounting fixtures and fasteners for proper fit and torque.

❑ Check the machine mountings that hold the machine to the floor or base block to make sure that they are secure.

❑ Check internal Morse tapers for burrs and scratches. Using an appropriate reamer, remove any raised metal that might interfere with operation.

Annual Maintenance

❑ Check all markings and safety information in and around the workstation. These should be clear, easy to read, and free of any defects that hinder visibility or legibility.

❑ Review and update safe operating procedures.

MEETS THE FOLLOWING (MSSC) PRODUCTION STANDARDS

P2-K3	Provide training to maintain equipment.
P2-A9C	Train others on the routine for maintaining equipment.
P7-K4D	Equipment is checked to ensure it is operating according to specifications.
P7-K4E	Tools are checked to ensure they are in compliance with specifications.
P7-A3B	Identify safety problems of aging equipment and replace or bring up to date.

Performing Maintenance on a Metal Lathe

The following general checklist outlines the basic operations and schedules for maintaining a metal lathe. This checklist is only a guide. It is not comprehensive. Check the owner's manual of your particular machine for a more detailed set of maintenance procedures.

Daily Maintenance

☐ Check that the appropriate personal protective equipment (PPE) is available and in good working order.

☐ Make sure that the workstation is clean and free of all obstacles and obstructions.

☐ Evaluate equipment to ensure that it is in good working order before the start of each day.

☐ Check equipment to ensure that it is operating according to specifications.

☐ Check that all safety guards are in place and functioning properly.

☐ Visually check all electrical interfaces such as switches and wiring. Arrange for the immediate repair of any detected problems such as damaged switches or cracks in wire sheathing.

☐ Check the manufacturer's specifications and lubricate the machine accordingly.

Weekly Maintenance

☐ Perform a thorough inspection of the gears, belts, and other internal mechanisms. They should be free of damage and operate smoothly.

☐ Test the gears, belts, and other internal mechanisms to ensure their proper operation.

☐ Lubricate the machine according to the manufacturer's specifications. Be certain to include the ways and slides.

Quarterly Maintenance

☐ Make sure that spare parts such as belts and cutting tools are available.

☐ Lubricate the machine according to the manufacturer's specifications.

☐ Adjust gib strips according to the manufacturer's specifications.

☐ Check for oil contamination in and around the workstation.

☐ If suds were used, replace them.

☐ Protect bare metal surfaces with a petroleum-based lubricant/solvent.

Six-Month Maintenance

☐ Clean waste from housings, cabinets, and other restricted access spaces.

☐ Check belts for flaws.

☐ Make sure the belts are set at the proper tension.

☐ Check pulley alignment.

☐ Check the mesh and alignment of the gear train.

☐ Examine all mounting fixtures and fasteners for proper fit and torque.

☐ Check the machine mountings that hold the machine to the floor or base block to make sure that they are secure.

☐ Check internal Morse tapers for burrs and scratches. Using an appropriate reamer, remove any raised metal that might interfere with operation.

Annual Maintenance

☐ Check all markings and safety information in and around the workstation. They should be clear and easy to read and free of any defects that hinder visibility or legibility.

☐ Review and update safe operating procedures.

MEETS THE FOLLOWING (MSSC) PRODUCTION STANDARDS

P2-A3D	Evaluate equipment to ensure that it is in good working order before the start of each day.
P7-K4D	Equipment is checked to ensure it is operating according to specifications.
P7-K4E	Tools are checked to ensure they are in compliance with specifications.
P7-A3B	Identify safety problems of aging equipment and replace or bring up to date.

COPYRIGHT © GLENCOE/McGRAW-HILL

Maintaining an Equipment Maintenance Log

An equipment maintenance log, also called a service and repair log, is a record of the maintenance and repair procedures performed on a particular piece of equipment. Such a log is essential in researching the reliability of equipment. The following are general guidelines for establishing and maintaining an equipment maintenance log.

❑ Store all equipment maintenance manuals in a single accessible location. Make sure they are arranged in a way that allows a new employee to easily locate the correct manual.

❑ Store the equipment maintenance log in an accessible location. These logs are sometimes maintained electronically. In such cases, the appropriate individuals should be able to use the computer to document the maintenance and repair history of a piece of equipment. Those needing information from the equipment maintenance log should have full instructions for accessing and maintaining that document. This may be necessary to determine maintenance and repair schedules and procedures. Those responsible for maintaining the log should be able to use the computer to enter data regarding scheduled maintenance and repair.

❑ Schedule a daily period for minor adjustments, cleanup, and lubrication.

❑ Facilitate agreement on the maintenance and repair schedule to minimize production impact.

❑ Make sure that maintenance and repair assignments are specific. If possible, assign the same individual or team to specific maintenance and repair operations. Practice and experience will generally help build skill levels.

❑ Be able to receive and understand maintenance and repair instructions.

❑ Read equipment manuals to determine proper maintenance and repair procedures.

❑ Review the equipment maintenance log to ensure that the recommended procedures are followed. Such review is essential to ensure proper equipment upkeep.

❑ Make sure that the equipment maintenance log is completed in a timely manner. It is important that all the information called for is provided.

❑ Return all maintenance manuals to their assigned location after use.

❑ If the equipment maintenance log is a print volume, return it to its assigned location after it has been updated. If the log is computerized, be sure that it has been properly saved.

The format of an equipment maintenance log is fairly simple. To assure consistency, it is important that all involved are in agreement on the format of the log. Most equipment maintenance logs call for the following information. Depending on procedures, most or all of this information should be provided by the technician performing the maintenance.

- Equipment.
- Building.
- Department.
- Serial number.
- Model number.
- Supervisor.
- Phone.
- Date tagged.
- Service/repair description.
- Parts ordered/Date ordered. Some facilities may have a deadline by which parts must be ordered.
- Date parts received.
- Date parts installed.
- Date repair completed.
- Repair technician's initials.

A sample page from an equipment maintenance log is shown on p. 178.

(continued on next page)

Equipment Maintenance Log						
Equipment			**Serial Number**			
Building			**Model Number**			
Department			**Supervisor**			
			Phone			

Person Performing Each Action Should Initial Within Box

Date Tagged	Service/Repair Description	Parts Ordered/ Date Ordered	Date Parts Received	Date Parts Installed	Date Repair Completed	Repair Technician's Initials

MEETS THE FOLLOWING (MSSC) PRODUCTION STANDARDS

P2-K1A	Preventive maintenance schedule is prepared and checked as appropriate.
P2-K1B	Preventive maintenance is performed to schedule.
P2-K1C	Preventive maintenance is documented completely and in a timely manner.
P2-K2E	Cross-training is provided when appropriate.
P2-O3B	Knowledge of forms and procedures for correctly documenting processes (e.g., preventive maintenance forms).
P2-O3D	Skill in documenting repairs, replacement parts, problems, and corrective actions to maintain log to determine patterns of operation.
P2-O3E	Skill in reviewing maintenance log/checklist to ensure that recommended preventive procedures are followed.
P2-A1B	Enter data into scheduled maintenance program using the computer.
P2-A1C	Use PC to document history of maintenance.
P2-A1D	Access documentation and log books electronically to determine maintenance schedules and procedures.
P2-A2C	Use maintenance instruction books and schedule logs to maintain machinery.
P2-A3C	Analyze machine repair logs to help determine the cause of equipment problems.
P2-A10E	Create agreement on the format of maintenance logs to ensure consistency.
P2-A13A	Receive maintenance instructions and understand them.
P2-A14D	Document maintenance and repair history of equipment.
P2-A15B	Read equipment manuals to determine proper preventive maintenance procedures, lubricants, and replacement parts.
P2-A15E	Review maintenance schedule in order to assure upkeep.
P7-O3B	Knowledge of equipment manual and standard practice manual to repair equipment safely.
P7-A3B	Identify safety problems of aging equipment and replace or bring up to date.
P7-A14C	Document how a piece of equipment complies with regulations.

COPYRIGHT © GLENCOE/McGRAW-HILL

Inspecting Powered Industrial Trucks

Powered industrial trucks (PITs) come in two general types: electric (battery) powered and internal-combustion engine (gas/LPG/diesel) powered. Each of these general types has a variety of different configurations and attachments.

Your workplace may have a variety of trucks, including:

- Electric forklift trucks.
- Propane forklift trucks.
- Yard forklift trucks.
- Electric transtackers.
- Riding grip tows.
- Stand-up riding tow tractors.
- Walking pallet trucks.
- Walking transtackers.
- Tow tractors.
- Industrial tractors.
- Reach trucks.
- Order pickers.

All truck operators must be trained to operate all of the types of PIT they will be operating. The OSHA standard for PIT training requires that an employer provide training to truck operators on a variety of topics. This includes vehicle inspection as well as the maintenance that the operator will be required to perform.

Each type of PIT is unique. Inspection checklists should apply to the specific vehicle being used. The manufacturer's instructions on vehicle maintenance and on the owner's and operator's responsibilities should also be consulted. The OSHA standards for PITs must be reviewed to ensure compliance.

OSHA requires daily pre-shift inspection of PITs. The supervisor is to be informed of any necessary repairs. Only a qualified mechanic can correct any problems. The following checklists serve only as a guide and may not be totally inclusive.

Sample Generic PIT Inspection Checklist

Consult the manufacturer's manual to learn which checks must be conducted with the engine or motor on and which must be conducted with the engine or motor off.

❑ **Overhead Guard**
- Are there broken welds, missing bolts, or damaged areas?

❑ **Hydraulic Cylinders**
- Is there leakage or damage on the lift, tilt, and attachment functions of the cylinders?

❑ **Mast Assembly**
- Are there broken welds, cracked or bent areas, and worn or missing stops?

❑ **Lift Chains and Roller**
- Is there wear, damage, kinks, signs of rust, or any sign that lubrication is required?
- Is there squeaking?

❑ **Forks**
- Are they cracked, bent, worn, or mismatched?
- Is there excessive oil or water on the forks?

❑ **Tires**
- Are there large cuts around the circumference of the tire?
- Are there large pieces of rubber missing or separated from the rim?
- Are there missing lugs?
- Is there bond separation that may cause slippage?

❑ **Battery Check**
- Are the cell caps and terminal covers in place?
- Are the cables missing insulation?

(continued on next page)

❑ **Hydraulic Fluid**
- Check the level.

❑ **Gauges**
- Check that they are working properly.

❑ **Steering**
- Is there excessive free play?
- If there is power steering, is the pump working?

❑ **Brakes**
- If the pedal goes all the way to the floor when you apply the service brake, that is the first indicator that the brakes are bad. Brakes should work in reverse, also.
- Does the parking brake work? The truck should not be capable of movement when the parking brake is engaged.

❑ **Lights**
- If the truck is equipped with lights, are they working properly?

❑ **Horn**
- Does the horn work?

❑ **Safety Seat**
- If the truck is equipped with a safety seat, is it working?

❑ **Load Handling Attachments**
- Is there hesitation when hoisting or lowering the forks, when using the forward or backward tilt, or in the lateral travel on the side shift?
- Is there excessive oil on the cylinders?

❑ **Propane Tank**
- Is the tank guard bracket properly positioned and locked down?

❑ **Propane Hose**
- Is it damaged? It should not be frayed, pinched, kinked, or bound in any way.
- Is the connector threaded on squarely and tightly?

❑ **Propane Odor**
- Check for the presence of propane gas odor. If you detect the presence of propane gas odor, turn off the tank valve and report the problem.

❑ **Engine Oil**
- Check levels.

❑ **Engine Coolant**
- Visually check the level. Note: Never remove the radiator cap to check the coolant level when the engine is running or while the engine is hot. Stand to the side and turn your face away. Always use a glove or rag to protect your hand.

❑ **Transmission Fluid**
- Check the level.

❑ **Windshield Wipers**
- Do they work properly?

❑ **Seat Belts**
- Do they work?

❑ **Safety Door** (found on stand-up rider models)
- Is the door in place?

❑ **Safety Switch** (found on stand-up riding tow tractors)
- Is the switch working?

❑ **Hand Guards** (found on stand-up riding tow tractors, walking pallet trucks, and walking transtackers)
- Are the guards in place?

❑ **Tow Hook**
- Does it engage and release smoothly?
- Does the safety catch work properly?

❑ **Control Lever**
- Does the lever operate properly?

❑ **Safety Interlock** (found on order pickers)
- If the gate is open, does the vehicle run?

❑ **Gripper Jaws** (found on order pickers)
- Do the jaws open and close quickly and smoothly?

(continued on next page)

COPYRIGHT © GLENCOE/MCGRAW-HILL

❑ **Work Platform** (found on order pickers)

- Does the platform raise and lower smoothly?

It is important that the truck operator read and understand the maintenance and safety procedures appropriate to the truck being operated. The operator should:

- Study the detailed operation procedures for equipment maintenance.
- Know how each individual piece of equipment functions.
- Make certain that the safeguards are in place.
- Read and understand the safety checks and operation procedures provided by the employer.

Once again, this checklist is not totally inclusive and is meant to be used only as a guideline. Refer to OSHA's Web site for more information on powered industrial trucks and their maintenance procedures.

Documentation

Besides maintaining checklists, operators must keep a record of:

- Fluids and fuel added for internal-combustion engine PITs.
- Fluids added for electric PITs.

See the following examples of record-keeping formats.

Record of Fluids and Fuel Added (for Internal-Combustion Engine PITs)	
Date	
Operator	
Department	
Shift	
Truck #	
Model #	
Serial #	
Hour Meter	
Fuel	
Engine Oil	
Radiator Coolant	
Hydraulic Oil	

(continued on next page)

Record of Fluids Added (for Electric PITs)	
Date	
Operator	
Department	
Shift	
Truck #	
Model #	
Serial #	
Drive Hour Meter Reading	
Battery Water	
Hydraulic Oil	
Hoist Hour Meter Reading	

MEETS THE FOLLOWING (MSSC) PRODUCTION STANDARDS

P2-04E	Skill in visually inspecting equipment to ensure safety compliance before operating.
P2-A3E	Visually inspect tools and equipment for possible wear and inform supervisor of needed repairs.
P2-A4A	Determine if equipment is operating properly.
P3-01J	Knowledge of machine functions to determine if all safeguards are operational.
P7-A14B	Create detailed safety operating procedures for equipment maintenance.
P7-A15A	Read and understand equipment maintenance safety procedures before operating equipment.

COPYRIGHT © GLENCOE/McGRAW-HILL

Testing Forklift Operator Performance

According to OSHA, employers must provide operation training and safety training for all those who will operate a powered industrial truck (PIT). Employers must certify that these operators have been trained and evaluated. The certification shall include the name of the operator, the date of the training, the date of the evaluation, and the identity of the person(s) performing the training or evaluation. See **6-9, Training PIT Operators**, for additional information about training.

One type of PIT is the forklift. Forklifts vary in design. Checklists for operator performance may vary. This sample forklift operator performance test can be adapted and expanded to cover the equipment being used. OSHA's requirements and the manufacturer's instructions on operator's responsibilities should be consulted. This checklist is only a guide regarding those actions on which the operator should be observed and evaluated. Before using this equipment or any other equipment in the workplace, check with your supervisor to see if you need specific training and certification.

Controls & Checks

- ❑ Showed familiarity with the controls.
- ❑ Followed proper instructions for maintenance checks at beginning and end of test.

Operation

- ❑ Checked bridge plates and ramps.
- ❑ Kept a clear view of direction of travel.
- ❑ Obeyed signs.
- ❑ Slowed down at intersections.
- ❑ Sounded horn at intersections.
- ❑ Gave proper signals when turning.
- ❑ Turned corners correctly and was aware of rear-end swing.
- ❑ Yielded to pedestrians.
- ❑ Drove under control and stayed within proper traffic aisles.
- ❑ Drove backward when required.

Load Handling

- ❑ Approached load properly.
- ❑ Checked load weight when parked, with controls neutralized, brake on, and power off.
- ❑ Ensured that load did not exceed load limit.
- ❑ Lifted load properly.
- ❑ Ensured that load was properly balanced.
- ❑ Ensured that forks under load all the way.
- ❑ Traveled with load at proper height.
- ❑ Carried parts or stock in approved containers.
- ❑ Maneuvered properly with load.
- ❑ Lowered load smoothly and slowly.
- ❑ Stopped smoothly and completely with load.
- ❑ Placed load within marked area.
- ❑ Stacked load evenly and neatly.

Additional Training

Refresher or remedial training would be required based on:

- Unsafe operation.
- An accident or near miss.
- Deficiencies found in a periodic evaluation of the operator.

Documenting Performance

It is important that all monitoring data is accurately documented. Maintain performance records according to company policy.

(continued on next page)

MEETS THE FOLLOWING (MSSC) PRODUCTION STANDARDS

P1-K4C	Operations are performed safely.
P4-A12C	Provide feedback on work performance that will maintain and improve performance.
P6-O3D	Knowledge of company performance evaluation policy and guidelines.
P7-K1G	Post-training evaluation indicates that workers can operate equipment safely.
P7-K4I	All monitoring data is accurately documented.
P7-O1A	Knowledge of government policies, procedures, and regulations governing the safe use of equipment.
P7-O1F	Knowledge of which tools and equipment require safety certification.
P7-O2A	Knowledge of equipment operation and design parameters to determine if machine is operating safely.
P7-A2C	Review safety requirements for a piece of equipment and integrate the requirements into the production procedures.
P7-A4A	Determine if equipment is being operated within appropriate safety standards.
P7-A9A	Practice safety in all areas, leading others by example.
P7-A17E	Apply environmental/safety and hazards standards in equipment operations.

COPYRIGHT © GLENCOE/McGRAW-HILL

Conducting a Job Safety Analysis

A job safety analysis (JSA) is a careful examination of safety practices to ensure the continued safety of all workers. You must follow the established procedure for conducting the analysis to ensure the accuracy and repeatability of the results. Note that the focus of a JSA is not on the performance of any given worker in a specific task. The focus is on the procedures of the task itself. By determining which of the given task's procedures is potentially harmful, it is possible to take corrective action to reduce or eliminate potential hazards. Once the JSA has been properly conducted, the results should be reviewed by the appropriate individuals. All changes needed to ensure the safety of the workers should be made. Each company will have its own method of conducting a JSA. The following is an example. See the JSA worksheet on p. 186.

1. **Select a job for analysis.** Remember that the purpose of the JSA is to review the job, not the workers performing the job. A JSA can be performed for every job, whether it is routine or otherwise. You might, for example, determine the adequacy of a safety plan for maintenance. In general, perform a JSA on:
   - Jobs with high rates of accidents and disabling injuries.
   - Jobs where close calls or near misses have occurred.
   - New jobs.
   - Jobs for which processes or procedures have been recently changed.

2. **Involve the workers.** When performing a JSA, you are evaluating a job, not the workers. Get input on potential hazards in job processes as well as ideas for correcting those hazards. Involve workers as much as possible when reviewing the job. They can often alert others to potential problems that might otherwise be overlooked.

3. **Conduct a preliminary JSA.** Observe the general conditions under which the job is performed. Make note of any obvious hazards.

4. **Break down the job.** List each step of the job in order of occurrence.

5. **Identify the hazards.** Study each step in detail. Record the hazards that currently exist or that may potentially occur.

6. **Evaluate the hazards.** Examine each of the hazards identified and determine what it is about the present job process that could create the hazard. What events might cause an injury?

7. **Recommend safe procedures and protection.** Evaluate the hazards identified with the workers performing the job and collaborate on ways to eliminate or reduce each hazard. Changing the job process, using safety gear, allowing more time, and upgrading equipment are all examples of ways to eliminate or reduce hazards.

8. **Revise the JSA.** Periodically review the JSA for any hazards that may have been overlooked previously. If an accident occurs, the JSA should be reviewed to determine if procedural changes are necessary.

9. **Document the analysis.** Note why changes were made and when they were made.

(continued on next page)

Job Safety Analysis Worksheet			Job to be Performed:	Tripping Pipe in Hole

Department: Drilling and Derrick Work **Date:** mm/dd/yyyy

JSA Completed by:	Task to be Performed by:		Supervisor
Full name	**Last**	**First**	Full name
	Last name	First name	

Personal Protective Equipment and Special Tools Required: Hardhat, safety-toe boots, safety glasses	**Supervisor Approval (signature)** *Supervisor's signature*

Step	Sequence of Basic Job Steps	Potential Hazards	Recommended Safe Job Procedures
1	Traveling block moving up derrick	Swinging blocks hitting sides of derrick. Tong counterweight line getting hooked on blocks or elevators.	Stabilize blocks and elevators. Do not put tongs on pipe too soon.
2	Put make-up tongs on and wrap spinning chain	Pinch points when latching tongs to pipe.	Keep hands and fingers on designated handles. Keep good tail on spinning chain. Keep control of chain.
3	Latching pipe into elevators	Pinch points of elevators and pipe. Dropping stand across derrick. Swinging pipe.	Derrickman should tail out pipe and stabilize stand after pickup. Floormen should watch for snag or short stand.
4	Stabbing pipe	Slipping while tailing pipe. Pinch points of pipe and tongs. Missing box.	Get firm hold. Give driller clear view. Place hands and legs properly.
5	Throwing chain, torquing pipe, unlatching tongs	Chain breaking, stuck by chain, pinch points—getting hand or fingers in chain. Tongs slipping.	Make sure tongs are latched properly. Hold tongs out of way after unlatching. Stay clear of chain and out of swing of tongs.
6	Pulling slips	Strains	Proper lifting techniques. Lift together. Use moving pipe as leverage.
7	Lowering pipe	Hitting bridge, line parting brake, or hydromatic failure.	Lower pipe at controlled speed. Watch weight indicator.
8	Set slips and unlatch elevators	Pinch points at slip handles, elevator links, and elevator latch.	Slow down pipe and set slips. All hands should work together. Proper lifting and hand placement.

MEETS THE FOLLOWING (MSSC) PRODUCTION STANDARDS

P3-A3D	Identify areas or tasks where most injuries occur to suggest modifications to process, layout, or job rotations in order to eliminate injuries.
P4-K3A	Potential improvements are generated through observation and data analysis.
P7-K3	Fulfill safety and health requirements for maintenance, installation and repair.
P7-K3B	Job safety analyses are reviewed regularly according to company policy.
P7-A4E	Determine adequacy of safety plan for maintenance.
P7-A5B	Organize time to ensure that safety operations are performed when needed.

   COPYRIGHT © GLENCOE/MCGRAW-HILL

Finding Present Value

A cost-benefit analysis considers all of the benefits of proposed alternatives. Some of these benefits may have a dollar value. In a cost-benefit analysis, you need to determine the "present value" of the future cost of the various benefits related to the alternative you are considering. To do this, you need to "discount" the future value of the cost to its present value. To determine present value, use the following formula.

$$PV = FV/(1 + r)^n$$

Where PV = Present Value

Where FV = Future Value

Where r = Discount Rate

Where n = Number of Years

The following example is simplified and does not consider depreciation, maintenance, or other factors. The example is designed to show you how to figure the present value of a future benefit that has a dollar value.

A certain change in production is estimated to save $100,000 five years from now. All of those savings will occur in the fifth year.

To bring about this change, you need to buy a specific production machine. You will need to replace this machine in five years. In order not to lose money on your investment in the machine, what is the maximum amount you should pay for the machine?

To determine the maximum amount to pay for the machine, you will need to determine the present value of the future savings. You can do this by using the following formula. You already know two of the items: Future Value (FV), which is $100,000, and Number of years (n), which is 5. The only item that is missing is the Discount Rate (r).

To use this formula, you need to provide a discount rate. The discount rate is an interest rate used to calculate the present values of expected future benefits and costs. You decide to use a discount rate of 10%. You select this rate because it is 1% less than the annualized return of the U.S. stock market over the past several decades.

PV = Present Value

FV = $100,000

r = 10

n = 5

Thus:

PV = $100,000/(1 + 0.10)^5

PV = $100,000/(1.61)

PV = $62,111

The formula indicates that the maximum price you should pay for the production machine is $62,111.

See **12-2, Performing a Cost-Benefit Analysis**, to use this formula.

MEETS THE FOLLOWING (MSSC) PRODUCTION STANDARDS	
P4-K3A	Potential improvements are generated through observation and data analysis.
P4-A7B	Show receptivity to alternative process methods that may improve productivity and reduce scrap.
P4-A16B	Perform cost/benefit analysis to determine if a CI idea is cost effective.
Core-A1K	**Making decisions and judgments:** Make decisions that consider relevant facts and information, potential risks and benefits, and short- and long-term consequences or alternatives.

Performing a Cost-Benefit Analysis

A cost-benefit analysis (CBA) is an evaluation of the costs and benefits of alternative approaches to a proposed activity. This analysis helps you determine the best alternative. There are 11 basic steps in performing a CBA.

1. **Determine and Define Objectives.** The individuals involved must understand what they are trying to accomplish. Include the project objectives and pertinent background information. Clearly define the problem. The CBA should be clear enough to be understood by a reviewer who is not intimately familiar with the organization and its work processes.

2. **Document the Current Process.** The baseline for any CBA is the current process. The CBA must thoroughly document the current process to ensure that everyone involved in the CBA preparation and review understands that process. The main areas to be documented are customer service, system capabilities, and system costs.

3. **Estimate Future Requirements.** Future production requirements affect system costs and benefits. Thus, it is very important to accurately estimate future requirements. The two key items to consider are the production system life cycle and the peak life cycle demands of the production system. In doing this, consider the following guidelines.

   - If possible, make more than one forecast using different estimating methods. This will check the original forecast. It will also add validity to the overall estimate.

   - Include production averages and peak demands in your estimates. If the system is not designed to meet peak demands, there must be a good reason (usually cost) not to do so.

   - Get feedback from other professionals on your estimates. They can point out potential shortcomings in the estimate or provide confirmation of methods and results.

   - Document everything. Documentation will facilitate updates to the estimate.

4. **Collect Cost Data.** Six sources of cost data are historical experience, current production system costs, market research, publications, analyst judgment, and special studies.

5. **Present Three Alternatives.** One alternative that should always be included is to continue with no change. The costs and benefits of that alternative may not have been documented.

6. **Document Assumptions.** If possible, justify assumptions on the basis of prior experiences or actual data. For example, you may assume that a certain item of production equipment will need to be upgraded every four years. This can also be an opportunity to explain why some alternatives were not included.

7. **Estimate Costs.** All costs for the full system life cycle for each alternative must be included. The following factors must be addressed: Activities and Resources, Cost Categories, Personnel Costs, Direct and Indirect Costs (Overhead), Depreciation, and Annual Costs.

8. **Estimate Benefits.** Benefits can be viewed as the return on investment (ROI). Benefits for customers are improvements to the current products or the addition of new products. Benefits for the manufacturer might include productivity gains or improved organizational effectiveness.

(continued on next page)

Copyright © Glencoe/McGraw-Hill

Estimate the value of the benefits. If a benefit cannot be assigned a monetary value, it might be assigned a relative numerical value. All benefits for the full system life cycle for each competing alternative must be included. Some examples of benefits for production systems are:

- **Accuracy.** Will the proposed alternative provide increased accuracy by reducing the number of production errors?
- **Availability.** How long will it take to develop and implement the system? Will one alternative be available sooner than another?
- **Compatibility.** How compatible is the proposed alternative with existing facilities and procedures? Will one alternative require less training of personnel or less new equipment?
- **Efficiency.** Will one alternative provide faster or more accurate processing of inputs? Will one alternative require fewer resources?
- **Maintainability.** Will the maintenance costs for one alternative be less than the others? Are the maintenance resources easier to acquire for one alternative?
- **Reliability.** Does one alternative provide greater reliability?

9. **Discount Costs and Benefits.** Find the present value of future benefits and costs. This is accomplished by discounting the future dollar value, which gives the present value of future benefits and costs. The present value (also referred to as the discounted value) of a future amount is calculated using the following formula:

$$PV = FV/(1 + r)^n$$

Where PV = Present Value

Where FV = Future Value

Where r = Discount Rate

Where n = Number of Years

10. **Evaluate Alternatives.** When the costs and benefits for each competing alternative have been discounted, rank the discounted present value of the competing alternatives. When the alternative with the lowest discounted cost provides the highest relative benefits, it is clearly the best alternative. If that is not the case, the evaluation is more complex. Most cases may not be that simple. Other techniques must then be used to determine the best alternative.

Some benefits may not have dollar values assigned. In this case, the non-cost values can be used as tie-breakers if the dollar-valued benefits do not show a clear winner and if the non-costed benefits are not key factors. If no benefits have dollar values, numerical values can be assigned (using a relative scale) to each benefit for each competing alternative. The evaluation and ranking are then completed on the basis of the number scores.

11. **Perform Sensitivity Analysis.** Sensitivity analysis identifies those inputs that have the greatest influence on the outcome. It repeats the analysis with different input parameter values and evaluates the results to determine which, if any, input parameters are sensitive. If a relatively small change in the value of an input parameter changes the alternative selected, then the analysis is considered to be sensitive to that parameter. If the value of a parameter has to be doubled before there is a change in the selected alternative, the analysis is not considered to be sensitive to that parameter.

(continued on next page)

APPLYING YOUR KNOWLEDGE

Compare and contrast worker hours of safe and unsafe practices for lifting heavy objects.

Procedure indicates that workers should wear a basic lifting belt provided by the company when lifting objects with dimensions larger than 2.5' × 2.5' × 2.5' and/or weighing more than 30 pounds. For all lifting operations, workers should squat down to the level of the object, grasp the object firmly with both hands, and use their legs—never their backs—to lift the object. Workers should stretch lightly before doing any lifting and again every 10 to 15 lifts. At the end of the shift or all lifting operations, whichever is sooner, the workers should perform a moderate to heavy stretch. At present, each worker averages 25 lifts in an eight-hour shift. Future requirements are likely to remain the same. Any increases or decreases in the number or nature of lifting operations is expected to be insignificant.

The alternatives are as follows:

Alternative 1: Keep everything as it is. The present safety system costs the company around $8,000 in equipment and $2,000 in lost productivity. It saves the company between $100,000 and $250,000 in costs due to injury.

Alternative 2: Improve the safety equipment. Improving safety equipment would add an extra $10,000 in equipment costs, bringing the yearly total to $20,000. The improved safety system is projected to save between $225,000 and $250,000 in costs due to injury.

Alternative 3: Remove all safety equipment and eliminate stretching. Eliminating the program could potentially cost the company as much as $250,000 per year in lost productivity due to worker absence and turnover as well as workers' compensation pay, medical bills, and lawsuits all due to injury. Eliminating the safety program would save the company $10,000 in equipment and productivity.

1. Given the above information, evaluate the three alternatives and rank them in order of their cost-to-benefit ratio. Which would be the best course of action for the company to take?

2. Calculate the present value (PV) of each of the alternatives. Use the lowest expected savings for the future value and assume a discount rate of 10%. Which alternative would allow the greatest investment based on its expected return? Which alternative would allow the least investment?

PRODUCTION CHALLENGE

You are on a work team at a petroleum refinery. One of the team's responsibilities is to manage their production budget. How would you work with your team to determine the cost-benefit analysis of adding overtime work to decrease overall production time?

MEETS THE FOLLOWING (MSSC) PRODUCTION STANDARDS

P4-K3A	Potential improvements are generated through observation and data analysis.
P4-A7B	Show receptivity to alternative process methods that may improve productivity and reduce scrap.
P4-A16B	Perform cost/benefit analysis to determine if a CI idea is cost effective.
P7-A16B	Compare and contrast man hours of safe to unsafe practices (ROI, CBA).
CORE-A1K	**Making decisions and judgments:** Make decisions that consider relevant facts and information, potential risks and benefits, and short- and long-term consequences or alternatives.

COPYRIGHT © GLENCOE/McGRAW-HILL

Managing Inventory

Accurate and reliable data are essential for efficient high-performance manufacturing. Inventory can represent a large part of the assets of a manufacturing facility. An inventory is made by physically counting the items. Managers and other decision makers need to know how much inventory there is and where it is located. This helps them make effective budgeting, operating, and financial decisions.

Managers also need to be informed of inventory discrepancies. An inventory discrepancy occurs when the inventory count does not match the number recorded as being on hand.

Inventory management is a process. This checklist identifies the key procedures.

1. Establish written policies.
   - ❏ Document the policies for the entire physical count process.
   - ❏ Regularly review and update established policies.
2. Develop written procedures for all aspects of the physical count process.
   - ❏ Define the individual tasks associated with the process.
   - ❏ Identify the procedures for completing required paperwork.
   - ❏ Provide examples of properly completed paperwork.
   - ❏ Regularly review established procedures.
3. Establish accountability.
   - ❏ Establish accountability and responsibility for the overall physical count.
   - ❏ Set inventory record accuracy goals at 95% or better.
   - ❏ Set other performance expectations.
4. Select an approach.
   - ❏ Use cycle counting to count all of the items in a certain class or physical location. The items that move the most quickly are counted the most frequently.
   - ❏ Use a wall-to-wall physical count to count every item in every location.

5. Determine the frequency of the counts.
   - ❏ Determine which items to count and how frequently to count them.
   - ❏ Choose a method of selecting individual items or locations for the count.
6. Enlist knowledgeable staff.
   - ❏ Ensure that counters are knowledgeable about the inventory items.
   - ❏ Ensure that counters are knowledgeable about the count process.
   - ❏ Ensure that counters are well trained.
7. Provide accurate supervision.
   - ❏ Assign members to count teams.
   - ❏ Assign team responsibilities.
   - ❏ Provide clear instructions.
   - ❏ Ensure that all items are counted.
   - ❏ Review count sheets.
8. Establish and maintain segregation of duties.
   - ❏ Use two-member count teams.
   - ❏ Provide proper supervision.
9. Perform blind counts.
   - ❏ Ensure that counters do not know how many of each item are recorded as being in inventory.
   - ❏ Do not give counters access to inventory system.
10. Execute the physical count.
    - ❏ Perform the requisite number of counts.
    - ❏ Verify item data and quantity.
    - ❏ Ensure that all items are counted.
    - ❏ Complete the counts in a timely manner.

(continued on next page)

11. Evaluate count results.
    - ❏ Compare the physical count with inventory records.
    - ❏ Adjust the inventory report by adding, subtracting, and dividing.
    - ❏ Identify storage practices leading to inventory damage, such as surface abrasion.
    - ❏ Modify policies and procedures to address needed changes in the physical count process.

12. Report count results.
    - ❏ Report count results to management, warehouse personnel, and counters.
    - ❏ Report count variances to management and security for approval and investigation.

MEETS THE FOLLOWING (MSSC) PRODUCTION STANDARDS

P1-K2C	Inventory discrepancies are communicated to the proper parties.
P1-A16A	Add, subtract and divide numbers to adjust inventory report.
P8-A3B	Identify storage practices and procedures to minimize surface abrasions on materials.

COPYRIGHT © GLENCOE/McGRAW-HILL

Completing a Repair Order Request

A repair order request is issued whenever an item of equipment is sent for repair or modification. A repair order request may be issued to a department within your company or to another company. The repair order request shown on p. 194 is typical. Note that this request has two sections. The person sending the equipment out for repair or servicing provides the information in the top section.

❑ **Issued To.** Provide the name of the department or company receiving the Repair Request Form.

❑ **Equipment Identification Number.** Provide the identification number or serial number of the item of equipment being repaired.

❑ **Order Number.** Enter the number assigned according to company procedures.

❑ **Date of Repair/Time In.** Enter the date and time that the item of equipment is sent for repair.

❑ **Authorized By.** Enter the name of the individual authorizing the repair.

❑ **Address.** Enter the street address, city, state, and ZIP code of the department or company performing the repair.

❑ **Equipment Manufacturer.** Enter the name of the manufacturer of the equipment being repaired.

❑ **Hour-Meter Reading.** If the item of equipment has an hour meter, enter the reading on the meter when the item is removed from service.

❑ **Date Completed/Time Out.** Enter this after the repaired item has been returned. Enter the date the repair was completed and the time on that day that the item was out of repair.

❑ **Downtime (Hours).** Enter the number of hours the equipment was out of service.

❑ **Description of Repairs or Services.** Describe the repairs or servicing required.

The repair technician will provide the information in the bottom section of the Repair Order Request. The technician will provide the following information for each part.

❑ **Date.** Enter the date the repair work was started.

❑ **Technician's Initials.** The technician performing the work should initial this column.

❑ **Labor Time.** Information in this column relates to Hours and the Hourly Rate for labor.

❑ **Hours.** Enter the actual number of hours (including fractions of an hour) needed to install the part or perform the service described in the column headed Part Number and Description.

❑ **Hourly Rate.** Enter the cost of labor per hour.

❑ **Labor Cost.** Multiply the number of Hours of Labor Time by the Hourly Rate. Enter the result.

❑ **Part Number/Description.** Enter a part number and description for each part used in the repair or servicing.

❑ **Quantity.** For each part, enter the number of parts used.

❑ **Unit Cost.** Enter the cost of each part used. Each individual part is a unit.

❑ **Parts Cost.** Multiply the Quantity by the Unit Cost. Enter the result.

❑ **Total Cost of Labor.** Add all of the costs in the Labor Cost column. Enter this total here. Enter this total also in Total Cost of Labor in the last column on the lower right.

❑ **Signature of Technician.** The technician performing the repair work should sign here.

❑ **Date Completed.** Enter the date the repair was completed.

❑ **Total Cost of Parts.** Add all of the costs in the Parts Cost column. Enter the total here.

❑ **Additional Cost, if Any.** Enter any costs other than those for Labor and Parts.

❑ **Total Cost.** Add the Total Cost of Parts, Total Cost of Labor, and Additional Cost, if Any. Enter the total here.

(continued on next page)

Repair Order Request

ISSUED TO		ADDRESS	
EQUIPMENT IDENTIFICATION NUMBER		EQUIPMENT MANUFACTURER	
ORDER NUMBER		HOUR-METER READING	
DATE OF REPAIR/TIME IN		DATE COMPLETED/TIME OUT	
AUTHORIZED BY		DOWNTIME (Hours)	

DESCRIPTION OF REPAIRS OR SERVICES

DATE	TECHNICIAN'S INITIALS	LABOR TIME		LABOR COST	PART NUMBER AND DESCRIPTION	QUANTITY	UNIT COST	PARTS COST
		HOURS	HOURLY RATE					

TOTAL COST OF LABOR	$	

I certify that the services and repairs noted herein have been completed.

TOTAL COST OF PARTS	$
TOTAL COST OF LABOR	$
ADDITIONAL COST, IF ANY	$
TOTAL COST	$

SIGNATURE OF TECHNICIAN — DATE COMPLETED

MEETS THE FOLLOWING (MSSC) PRODUCTION STANDARDS

P2-K1D	Repair needs are communicated to the correct parties using the right procedures and forms.
P2-O3D	Skill in documenting repairs, replacement parts, problems and corrective actions to maintain log to determine patterns of operation.
P2-A14A	Write out repair order requests.

COPYRIGHT © GLENCOE/McGRAW-HILL

Scheduling Production

Production planning is concerned with establishing the routes and schedules for manufacturing the product. The plan for processing the materials in the plant involves the functions of routing, loading, scheduling, and dispatching. A purchasing department is responsible for obtaining the supplies and materials needed to make the product.

- Routing is the function of determining the route the material must take through the plant.
- Loading is the function of calculating the time required to process a job lot and charging it against available production time.
- Scheduling is planning when the work is to be done.
- Dispatching is authorizing the start of a manufacturing operation.

After routing is completed, loading and scheduling can begin. Loading and scheduling are often done on special charts. These charts provide at-a-glance simplicity to the complexity of planning.

Loading. A loading chart consists of a column listing specific machines or workstations. The header row contains titles for time periods (e.g., weeks). A light line in one of the cells usually indicates a commitment for a particular time period and a dark line usually indicates a backlog of work.

Scheduling. A scheduling chart is similar to a loading chart. A column on the left lists the machines or manufacturing processes. A row along the top represents the dates. Brackets are used to indicate the start/stop times for each machine or process. Progress is tracked by drawing a line between the brackets that is in proportion to the amount of work completed.

Examine the charts and use them to:
- ❑ Make job assignments.
- ❑ Ensure that job assignments match skills with work to be done.
- ❑ Maximize the use of available skills.
- ❑ Ensure that business and customer needs are met.
- ❑ Verbally clarify customer needs to co-workers.
- ❑ Achieve production quotas.
- ❑ Effectively notify workers of assignments.
- ❑ Notify team members of schedule requirements in a timely way.
- ❑ Make job assignments and coordinate workflow with team members and other work groups.
- ❑ Present concerns to a supervisor about production schedules and personnel needed to meet the schedule.
- ❑ Schedule workers with appropriate skills according to production needs.
- ❑ Estimate production time to determine and clearly communicate delivery schedules and cost.
- ❑ Analyze the production schedule and assign work-related duties.
- ❑ Work with all team members to coordinate material flow across multiple processes or workstations to assure on-time delivery.
- ❑ Check tools and equipment against the work order.
- ❑ Coordinate inter-department requirements to ensure that the final product meets specs.
- ❑ Determine availability of equipment retooling to fulfill production requirements.
- ❑ Facilitate agreement on job assignments in order to meet requirements and schedules.
- ❑ Change schedules to adapt to production needs without sacrificing equipment efficiency.
- ❑ Calculate time estimates for jobs and schedules for goals.
- ❑ Communicate product availability to the proper parties in a timely way.

(continued on next page)

If a particular workstation needs only two operators to run efficiently, and a new product suddenly requires a new process, pull workers from stations where they can be spared and reassign them to work on the new process. In this way, the production schedule is adapted to changing production needs without sacrificing the efficiency of any given workstation. These charts are also useful in preplanning production. It is possible to determine the availability of equipment for production as well as estimate the time to completion so delivery schedules and costs can be figured.

Creating Production Schedules

Create two production schedule charts for the production of a part. One chart will schedule the activity of the individual workstations. The other will track the progress of the whole process. Note the following.

- There are three mills and three lathes available.
- Each part requires three cuts in a mill and three cuts in a lathe.
- It takes one machine one week to make one cut in all of the parts. Each additional cut doubles the amount of time required.
- The order of the cuts does not matter. (Lathe cuts can come before mill cuts or vice versa.)

- The entire process from design to completed product can be no more than six months.
- Engineering will take at least two months with no setbacks.
- Setup and procurement of materials are two separate processes that will each take two months. They may be performed simultaneously with each other and with engineering. However, setbacks in engineering will delay setup and procurement and possibly extend the required amount of time for each beyond two months. A one-month lead for engineering is recommended to prevent problems.
- The time it takes to manufacture the product will depend upon the schedule determined earlier. It cannot overlap engineering. It may overlap setup and procurement by one month, but only one process can be running during the overlap.
- Packaging and shipping will take one month and cannot overlap manufacturing.

Schedule the manufacture of the parts and the whole production process to minimize downtime and complete the project on time or early. See the possible solution on p. 197.

(continued on next page)

COPYRIGHT © GLENCOE/MCGRAW-HILL

Possible Solution

Machine Tool Scheduling Chart

	Days									
	1	2	3	4	5	6	7	8	9	10
1st half of parts					Changeover: mill parts go to lathes and lathe parts go to mills, 4.7 days					Product completion: 9.3 days
Mill 1	▬▬▬▬▬▬▬▬▬									
Mill 2	▬▬▬▬▬▬▬▬▬									
Mill 3	▬▬▬▬▬▬▬▬▬									
2nd half of parts										
Lathe 1	▬▬▬▬▬▬▬▬▬									
Lathe 2	▬▬▬▬▬▬▬▬▬									
Lathe 3	▬▬▬▬▬▬▬▬▬									

Manufacturing Process Scheduling Chart

	Months					
	1	2	3	4	5	6
Engineering	▬▬▬▬	▬▬				
Setup		▬▬▬	▬			
Procurement		▬▬▬	▬			
Manufacturing				▬		
Packaging and Shipping				▬▬		

(continued on next page)

COPYRIGHT © GLENCOE/MCGRAW-HILL

MEETS THE FOLLOWING (MSSC) PRODUCTION STANDARDS

P1-K2B	Tools and equipment are checked against work order.
P1-K2E	Workers with appropriate skills are scheduled according to production needs.
P1-K7C	Product availability is communicated to the proper parties in a timely way.
P1-O4N	Knowledge of how to estimate time to determine delivery schedules and cost.
P1-A8A	Coordinate inter-department requirements to ensure final product meets specs.
P1-A8B	Work with all team members to coordinate material flow across multiple processes or workstations to assure on time delivery.
P1-A12A	Verbally clarify customer needs to co-workers.
P1-A12B	Present concerns to supervisor about production schedules and personnel needed to meet that schedule.
P2-A4B	Determine availability of equipment retooling to fulfill production requirements.
P2-A7D	Change schedules to adapt to production needs, while not sacrificing equipment efficiency.
P5-K2	Communicate material specifications and delivery schedules.
P5-K2B	Delivery schedules are clearly communicated.
P5-O2A	Skill in calculating time estimates for jobs.
P6-K3	Make job assignments.
P6-K3A	Job assignments match skills with the work to be done.
P6-K3B	Job assignments maximize the use of available skills.
P6-K3C	Job assignments ensure business and customer needs are met.
P6-K3D	Workers are notified of assignments effectively.
P6-K4	Coordinate work flow with team members and other work groups.
P6-K4B	Team members are notified of schedule requirements in a timely way.
P6-O1A	Skill in making job assignments and coordinating workflow.
P6-A2D	Analyze production schedule to assign work-related duties.
P6-A10D	Facilitate agreement on job assignments in order to meet requirements and schedules.
P6-A14B	Fill out production schedule to achieve production quotas.
P6-A16A	Calculate schedule goals.
P6-A16C	Analyze production data to determine manpower requirements needed to achieve goals.

   COPYRIGHT © GLENCOE/McGRAW-HILL

Completing a Material Requisition Form

A materials requisition form is an order form used to request needed materials. These forms are completed as print items. They can also sometimes be completed on-line. Such communications should be tracked and documented as appropriate. To ensure prompt delivery of requested items, you must properly complete the materials requisition form. The sample form shown on p. 200 contains the main elements present in all requisition forms. Some materials requisition forms are more detailed than this one.

- **Date.** Insert the date the requisition form was filled out.
- **Requisition Number.** This number is required for tracking purposes. Follow the preferred requisition numbering method used in your facility.
- **Recommended Vendor.** List the known vendor's name and address, as well as their telephone number and fax number. If unknown, enter Best Source.
- **Location.** Print your location on this line.
- **Ship To Address.** Print the address to which the requisition is to be shipped.
- **Special Instructions.** Note a Rush Order here. The notation ASAP (As Soon As Possible) is a general request for special handling. If you require the items by a certain date, specify that date here.

- **Quantity.** The quantity relates to the Unit of Issue.
- **Item Description.** Briefly describe the item.
- **Unit of Issue.** Indicate the Unit of Issue (each, box, set, etc.).
- **Unit Cost.** The Unit Cost must agree with and relate to the same Unit of Issue entered in Quantity.
- **Total Cost.** This should equal the Quantity multiplied by the Unit Cost.
- **Requisition Total.** This is the total amount of the requisition.
- **Approval Signature.** The form is to carry the signature of the appropriate person. Generally this is a person with expenditure authority.

MEETS THE FOLLOWING (MSSC) PRODUCTION STANDARDS

P1-O4D	Knowledge of how to order materials and tools.
P1-A14C	Fill out order form.
P5-K2F	Communications are tracked and documented, as appropriate.
P5-A14A	Complete a material requisition form when parts are needed.

(continued on next page)

Material Requisition Form

To: Date:_____

☐ Purchasing Department   Requisition Number:_____
   (for purchase order)

☐ Accounts Payable
   (for check writing)

Recommended Vendor:_____   Location:_____
_____   _____
_____   _____
_____   _____

Ship to:_____   Special Instructions:_____
_____   _____
_____   _____
_____   _____

Quantity	Item Description	Unit of Issue	Unit Cost	Total Cost
			Requisition Total	

Approval Signature:_____

Completing an Inventory Report Form

Every manufacturing process requires an inventory of materials. Even just-in-time processes, in which the material on hand is kept at the lowest possible level, require some inventory. Inventory reports record the identity and amount of materials held in inventory or in the process. Because there are costs associated with having too much or too little material on hand, accurate records are necessary. These records guide decisions to:

- Place or modify an order.
- Move materials from a warehouse.
- Schedule production.

An inaccurate inventory record can lead to costly errors, so it is important to enter all information into an inventory record carefully. In particular, pay attention to mathematical calculations, a common source of errors.

Entering & Calculating Inventory Information

There is no standard inventory form. The format of the inventory report depends on the company's needs and often varies within an operation, depending on how the information is used. The following tables show portions of typical forms used to track the amount of material on hand. Note that the final column of this report is a calculated result, based on addition and subtraction of other columns. The units received are added to the initial units on hand. For example, for part number A-28-L, 38 + 15 = 53. The units transferred to production are then subtracted from that figure, 53 − 29 = 24. If the report involves calculated numbers, the results should always be double-checked to be sure that the final number is reasonable.

Weekly Inventory of Production Parts				
Part number	Initial units on hand	Units received	Units transferred to production	Final units on hand
A-28-L	38	15	29	24
A-28-R	42	15	29	28
A-29-L	73	0	29	44
A-29-R	28	1	29	0

When you record data into an inventory form, pay particular attention to the units being recorded. While some data may be based on a count of individual units, other information could be recorded as a number of packages or cases. Entering the number of cases where the number of units is required can cause large discrepancies in the report.

Total Parts on Hand				
Part number	Number of cases	Units per case	Additional units	Total units on hand
A-28-L	10	25	3	253
A-28-R	11	25	5	280
A-29-L	15	12	0	180
A-29-R	8	12	10	106

(continued on next page)

APPLYING YOUR KNOWLEDGE

1. Examine this inventory report and identify two errors in the final units on hand column.

Weekly Inventory of Production Parts				
Part number	Initial units on hand	Units received	Units transferred to production	Final units on hand
transmissions	28	15	29	24
clutches	68	12	30	50
pressure plates	17	18	6	39
bearings	28	1	29	0

2. Complete this inventory report.

Total Parts on Hand				
Part number	Number of cases	Units per case	Additional units	Total units on hand
Number 8 screws	7	100	78	
Number 10 screws	13	100	0	
Number 12 screws	8	250	346	
Number 14 screws	20	250	18	

3. The standard operating procedure requires that 250 gears be in inventory. A warehouse count indicates the inventory has dropped to 75 units. How many cases of 25 gears should be ordered?

PRODUCTION CHALLENGE

The manufacture of baseball gloves includes producing the individual parts of the glove and then sewing them together. These parts include the shell, the lining, the pad, and the web, all made of leather. Plastic reinforcements and rawhide lacings are also used. Your work includes keeping track of the inventory of these parts to make sure enough are on hand when the company receives orders. How would you set up an inventory procedure?

MEETS THE FOLLOWING (MSSC) PRODUCTION STANDARD

P1-A16A	Add, subtract and divide numbers to adjust inventory report.

COPYRIGHT © GLENCOE/McGRAW-HILL

Storing Materials

Proper storage of materials depends on attention to the main objectives of storage. The main objectives of materials storage are:

- **Protection.** Materials must be protected from all risk of damage, including surface abrasion. To prevent damage, materials should not be roughly handled. Employees engaged in warehouse and storage operations must be instructed in safety and fire protection regulations pertaining to these operations. Storage buildings must meet building and fire code requirements. Storage areas must be inspected regularly for leaks or spills.

- **Accessibility.** Materials must be easily accessible.

- **Order.** Similar materials must be stored in one area.

- **Identification.** Materials must be clearly identified.

- **Rotation.** Material shelf life must be known. Inventory must be managed on a first-in-first-out basis.

Storage methods will vary according to available space, labor, equipment, and the quantity and type of material. Identification of materials is essential. The following checklist provides examples of basic methods of storage and identification.

Material in Bins

❑ Label each bin with a full description of the bin contents. Use identifying names on bin labels or tags that are identical to those on the warehouse inventory documents.

❑ Where two or more different items are stored in a single bin, identify the items by tagging one of each item.

Original Packages

❑ Label packages to indicate contents.

Metal Pipe & Steel

❑ Identify items by size and type on a tag or bin card on each storage rack. A color code may also be used to identify steel by class.

Large Items of Hardware

❑ Tag large items separately to show size and type.

Lumber

❑ Mark each stack by size, grade, and kind.

Prefabricated Units Shipped Unassembled

❑ Tag each component part or container. The tag should carry the item number assigned to the prefabricated unit.

Salvaged & Used Materials Returned to Stock

❑ Tag individual pieces or containers to show manufacturer's part number, description, and machine for which used. Identify as "used."

Equipment Spare Parts

❑ Tag spare parts with the part name, manufacturer and part number, equipment identification, and specification number. Protect tag with envelope to prevent fading.

Other Materials & Supplies

❑ Mark all other materials and supplies so that they may be readily identified without reference to other records.

(continued on next page)

Hazardous Materials

Hazardous materials require special treatment in storage. Specified storage practices must be followed to minimize the risk of environmental contamination and fire, including spontaneous combustion. The following is a general checklist of storage practices for hazardous materials. More information, including information on the storage of specific hazardous materials, can be obtained from the Occupational Safety and Health Administration (OSHA).

❏ Make sure that a Material Safety Data Sheet (MSDS) is readily available for each hazardous substance.

❏ Label each container for a hazardous substance with the product identity and a hazard warning.

❏ Use approved containers and tanks for the storage and handling of flammable and combustible liquids.

❏ Make sure that storage rooms for flammable and combustible liquids have explosion-proof lights.

❏ Make sure that storage rooms for flammable and combustible liquids have mechanical or gravity ventilation.

❏ Place firm separators between containers of combustibles or flammables, when stacked one upon another, to assure their support and stability.

❏ Separate fuel gas cylinders and oxygen cylinders by distance and fire-resistant barriers while in storage.

❏ Store liquefied petroleum gas in accordance with safe practices and standards.

❏ Post NO SMOKING signs on liquefied petroleum gas tanks.

❏ Make sure that liquefied petroleum storage tanks are guarded to prevent damage from vehicles.

❏ Make sure that storage tanks are adequately vented to prevent the development of excessive vacuum or pressure as a result of filling, emptying, or atmosphere temperature changes.

❏ Make sure that storage tanks are equipped with emergency venting that will relieve excessive internal pressure caused by fire exposure.

❏ Make sure that bulk drums of flammable liquids are grounded and bonded to containers during dispensing.

❏ Make sure that connections on drums and combustible liquid piping are vapor and liquid tight.

MEETS THE FOLLOWING PRODUCTION STANDARDS

P2-K4B	Materials are kept in a safe manner.
P8-A3B	Identify storage practices and procedures to minimize surface abrasions on materials.

COPYRIGHT © GLENCOE/MCGRAW-HILL

Calculating Downtime

Downtime is time during which production is stopped. Often this is for setup, maintenance, or repair. Because downtime represents a financial loss to the manufacturer, it must be minimized. There are two types of downtime: scheduled downtime and unscheduled downtime.

Scheduled Downtime. Scheduled downtime is downtime that is a planned part of the production schedule. It has the following characteristics:

- It can include regular maintenance, process changes, and similar operations.
- It can include shift changes.
- It can often help eliminate unscheduled downtime by addressing problems before they become critical.
- Its cost can be anticipated.

Unscheduled Downtime. Unscheduled downtime is downtime that is not a planned part of the production schedule. It has the following characteristics:

- It often results from the failure of a critical piece of equipment.
- It sometimes results from a poorly implemented maintenance schedule.
- It is usually more costly than scheduled downtime.

The possibility of unscheduled downtime should be reduced as much as possible. Input from machine operators can often help reduce unscheduled downtime. They can help identify the part or the operation that is causing the downtime. Their input can also be used to reduce or eliminate such downtime.

There is no standardized way of calculating the cost of downtime. Each industry has its own method of accounting for downtime. Companies in the same industry may calculate the cost differently. Despite this, there are some general calculations that can be used to estimate downtime costs.

1. Calculate the number of downtime hours.
2. Calculate the cost of hourly downtime labor.
3. Calculate the cost of product sales lost due to downtime.
4. Calculate the total cost of the necessary repairs including parts, labor, and any shipping and handling costs.
5. Calculate the start-up costs.
6. Add all of the costs to obtain the total cost.
7. Divide by the number of downtime hours to obtain the hourly cost of downtime.

The following example indicates how much downtime can cost in terms of production costs and lost sales. The downtime calculation formula will help you figure the cost of downtime in a workplace situation.

A manufacturer operates at 90% efficiency and produces 9,000 gears per hour. Each machine is operated by two skilled machinists who each make $35.00 an hour. Recently, one of the 10 machines used in manufacturing the gears broke down unexpectedly. The main drive shaft on the machine wore out and the machine shut down. Consider the following costs.

- The two workers, still being paid at their hourly rate, were left with nothing to do until the machine was repaired.
- A repair crew was contracted to come into the factory to assess and fix the problem. Emergency repairs are billed at $500 per hour plus the cost of parts.
- The drive shaft is a specialty part that must be custom manufactured. The cost for rush manufacture of the part is $20,000, twice the normal cost of the part.
- Rush shipment of the part to the repair crew costs $1,000 due to its size and the distance it is shipped.
- The manufacturer can expect $10 per part in gross revenue. Each machine produces an average of 900 parts per hour for a total expected income of $9000 per hour.

(continued on next page)

COPYRIGHT © GLENCOE/McGRAW-HILL

MANUFACTURING APPLICATIONS **205**

When the machine shuts down, the manufacturer contacts the repair crew. The following sequence of events then occurs.

1. The repair crew takes one hour to reach the manufacturer and one hour to diagnose the problem.

2. After diagnosing the problem, the repair crew spends another hour locating a supplier for the part. The part will take no less than seven hours to produce, an hour to package, and three hours to ship.

3. The repair crew then spends four hours disassembling the machine to prepare it to take the new part. They discover that the driveshaft failure also damaged several gears and other assorted parts. They order the parts for a sum total of $1500. They receive them and prepare them to go into the machine. The repair crew then waits for the delivery of the drive shaft.

4. Two hours later the supplier calls to say that the part is on its way and should arrive in three hours.

5. An hour later the first shift ends. The two machinists are replaced by the second-shift machinists, who make $52.50 an hour.

6. Two hours later, the drive shaft arrives.

7. It takes one hour for the repair crew to unpack the part and another hour for them to seat the part.

8. It then takes five hours to reassemble the machine, one hour to conduct the necessary safety checks, and one hour for the machine to be back in production.

Following is a breakdown of the cost of this unscheduled downtime of 23 hours.

- Employee labor: $997.50.
- Repair crew: $10,500.00.
- Parts, including shipping: $22,500.00.
- Loss of gross income due to nonproduction of gears: $207,000.00.
- Total cost: $240,997.50.

Though this example may seem extreme, it demonstrates that downtime can be costly. Indeed, just one more hour of downtime in this example would have pushed the total cost over a quarter of a million dollars. This is also just one downtime incident. Four or five similar incidents in a year could easily push the downtime cost above one million dollars.

MEETS THE FOLLOWING PRODUCTION STANDARDS

P2-A13E	Listen to machine operator in order to understand what part of the process is causing the most downtime.
P2-A16E	Calculate scheduled downtimes for machine maintenance.
P5-A16D	Calculate downtime caused by part shortage.

    COPYRIGHT © GLENCOE/McGRAW-HILL

Performing Preventive Maintenance

Preventive maintenance (PM) consists of inspections, adjustments, and routine repairs performed to prevent equipment malfunction. You can influence others to undertake a PM program. The following operations are general guidelines for performing PM on a CNC machine tool. Always refer to the manufacturer's equipment manual for PM procedures and schedules.

General PM Procedures

PM must be properly performed whether on a daily, weekly, monthly, or periodic basis. The following are general guidelines.

❑ Ensure that PM training materials are documented and available.

❑ Prepare the PM schedule by referring to the manufacturer's equipment manual.

❑ Estimate the repair or maintenance time to determine if the work will impact the production schedule.

❑ Develop checklists of inspections, adjustments, lubrications, and routine repairs.

❑ Ensure that the appropriate tools and supplies are on hand for PM.

❑ Make sure that PM is performed by qualified personnel skilled in following PM schedules.

❑ Monitor equipment indicators to ensure that it is operating correctly.

❑ Compare current equipment performance to optimal equipment operations.

❑ Make any adjustments correctly to ensure that the equipment is operating within established parameters.

❑ Apply statistical analysis to PM operations to help with reliability tracking, failure trend analysis, and measuring maintenance effectiveness.

❑ Review the PM log or checklist to ensure that recommended measures are followed.

❑ Follow up on any repair work. Estimate equipment performance after maintenance or repairs are completed.

❑ Document PM procedures in a timely manner according to company policy. This will help ensure that the equipment repair history is complete, up-to-date, and accurate. A review of PM documents can help forecast repair costs and evaluate equipment reliability.

Daily PM Procedures

❑ Clean machine tool.

❑ Inspect oil levels and correct if necessary.

❑ Inspect air-supply dryer water trap.

❑ Inspect power drive belts.

❑ Inspect coolant levels and correct if necessary.

❑ Inspect machine guards.

❑ Turn on machine and check for warnings and alarms on control panel. Correct before proceeding.

❑ Document inspection in a timely manner according to company policy.

Weekly/Monthly PM Procedures

❑ Clean machine tool.

❑ Inspect air filters to ensure that they are clean and free of obstruction.

❑ Check coolant mixture concentration.

❑ Check air supply line pressures.

❑ Check regulator pressures.

❑ Lubricate components requiring manual lubrication.

❑ Inspect machine guards.

❑ Turn on the machine and check for warnings or alarms on control panel. Correct before proceeding.

❑ Document inspection in a timely manner according to company policy.

(continued on next page)

Periodic PM Procedures

Periodic PM is performed according to the schedule set by the manufacturer. Some of the following tasks may require special tools or specific job experience. These tasks should be performed by qualified personnel.

❑ Clean machine tool.

❑ Check oil level in gearboxes and housings.

❑ Inspect belts for looseness and wear.

❑ Inspect pulleys for secure mounting and wear.

❑ Inspect interior of control unit cabinet.

❑ Inspect way guards.

❑ Inspect way wipers.

❑ Change coolant.

❑ Ensure that all safety devices are in place and functional.

❑ Power up the machine and test for warning codes and alarms. Correct before proceeding.

❑ Perform basic tool operations using the basic operations of machine controls and machine axes.

❑ Document inspection in a timely manner according to company policy.

MEETS THE FOLLOWING (MSSC) PRODUCTION STANDARDS

P2-K1	Perform preventive maintenance and routine repair.
P2-K1A	Preventive maintenance schedule is prepared and checked as appropriate.
P2-K1B	Preventive maintenance is performed to schedule.
P2-K1C	Preventive maintenance is documented completely and in a timely manner.
P2-K1E	Any necessary repair work is checked through follow-up.
P2-K1F	Necessary supplies are available to do the preventive maintenance.
P2-K2	Monitor equipment indicators to ensure it is operating correctly.
P2-K2A	Current equipment performance is regularly compared to optimal equipment operations.
P2-K2B	Abnormal equipment conditions are investigated.
P2-K2C	Abnormal equipment conditions are corrected in a timely manner.
P2-K2E	Documentation of equipment repair history is complete, up-to-date, and accurate.
P2-K3B	Preventive maintenance training materials are documented and available.
P2-O1C	Skill in following preventive maintenance schedules.
P2-O1E	Skill in repairing and maintaining machines or tools.
P2-O1G	Skill in recognizing wear and tear on equipment components.
P2-O2B	Skill in using appropriate maintenance tools to maintain machines.
P2-O3E	Skill in reviewing maintenance log/checklist to ensure that recommended preventive measures are followed.
P2-A4C	Determine preventive maintenance schedule in accordance with production schedule.
P2-A4D	Determine if equipment maintenance will impact production schedule.
P2-A9D	Influence others to perform preventive maintenance and repairs.
P2-A16A	Determine volume of coolant and oils.
P2-A16B	Measure liquid quantities.
P2-A16C	Estimate repair time for equipment.
P7-K4H	Equipment and process are adjusted correctly.
P7-O2C	Skill in making adjustments to equipment to ensure that is operating within established parameters.
P7-A16C	Apply statistical analysis to preventive maintenance operations.
P7-A16D	Estimate performance of equipment after repairs are completed.

COPYRIGHT © GLENCOE/MCGRAW-HILL

Calculating Productivity Levels

A manufacturer's profits are based in large part on its productivity levels. Productivity must align with demand for the product. If products are not available to customers when they want them, they might find another source. The manufacturer will then lose sales. If too many product units are produced in excess of immediate demand, the manufacturer incurs storage and management costs. These costs affect profits too.

Productivity is a measure of performance over a period of time, including downtime. It is expressed as a percentage.

To calculate productivity, first find the standard rate. It is usually determined by using data from past production cycles for that product. Take the average number of hours actually spent producing the product. Divide that by the average number of units produced in that time. The result is the standard production rate in hours per unit. It is usually a fraction of an hour. For example, in a production period of 10 hours, 100 units are produced. Dividing the hours (10) by the number of units (100), the standard rate is 0.1 hours.

Calculating Productivity

1. Determine the standard rate as described above.

2. Determine the standard hours by multiplying the standard rate by the number of parts produced in a given shift.

3. Determine the total time for a given shift, including downtime.

4. Calculate the productivity. Divide the standard hours by the total time for the shift and multiply that number by 100 to get a percentage value.

> **Productivity formula:** [(number of parts × standard rate) ÷ total hours] × 100 = productivity (as a percentage)

Sample Problems

1. An operator produces 101 parts in a 12-hour shift. Each part has a standard rate of 0.0935 hours/part. What is the productivity level? **Answer: 78.7%**

2. If a production worker's productivity level is 97.85% and she produces 76 parts over her 8-hour shift, what is her standard rate of production in hours? **Answer: 0.103**

Various factors affect productivity level. Ideally, it should be at 100%. This means that the exact necessary amount of product is produced in the exact necessary amount of time with no downtime. However, underproduction and overproduction occur in real manufacturing settings. On the whole, productivity should average between 90% and 110%. If productivity levels are too low, follow these suggestions to bring them back on track.

- Change schedules to adapt to production needs without sacrificing equipment efficiency.

- Improve efficiency at your workstation and suggest ways for co-workers to do the same.

- Reduce or eliminate downtime due to preventable causes.

MEETS THE FOLLOWING (MSSC) PRODUCTION STANDARDS

P2-A7D	Change schedules to adapt to production needs, while not sacrificing equipment efficiency.
P2-A12A	Suggest how a co-worker can improve workstation efficiency.
P6-K4D	Downtime is minimized.
P6-A16B	Calculate team productivity levels.

Preparing Standard Operating Procedures

A standard operating procedure (SOP) is a set of written instructions that document an activity. An SOP may be written for any routine or repetitive technical or administrative procedure. An SOP may be needed even when published instructions are available to provide greater detail. SOPs are essential in successful quality assurance plans and safety programs.

SOPs provide individuals with the information needed to perform a job safely and correctly. By documenting the way activities are to be performed, they help ensure safe work practices and consistent product quality. SOPs also:

- Minimize miscommunication.
- Reduce work effort.
- Can be used in a personnel training program, because they provide detailed work instructions.
- Can be used to reconstruct project activities.
- Are frequently used as checklists by inspectors auditing procedures.

Writing

The company should have a procedure for identifying processes that need to be documented by an SOP. They should also have in place a procedure for preparing such an SOP. If not prepared correctly, an SOP is of limited value. Preparation of an SOP should follow these general guidelines.

- The writer should be someone with extensive knowledge of the activity. This would be the production worker who actually performs the work or uses the process.
- For a multitasked process requiring a team approach, several individuals would contribute to the SOP.
- The work experience needed to perform the activity should be noted in the section on personnel qualifications. The need for any prior required training should be noted.

General Format

SOPs should be organized for ease of use. There is no single correct format. In general, technical SOPs have five main parts, in the following order. This format can be altered as required.

- Title page.
- Table of contents.
- Procedures.
- Quality assurance and quality control.
- References.

Title Page. The title page is the first page of the SOP. It should contain:

- A title that clearly identifies the activity or procedure.
- An SOP identification number.
- Date of issue and/or revision.
- The name of the department to which the SOP applies.
- The page number (in the format: Page 1 of X).
- The signatures and signature dates of those who prepared and approved the SOP. Electronic signatures are acceptable for SOPs maintained on a computer database.

Table of Contents. The table of contents should list the sections with their page numbers. It should also note changes and revisions made to those sections.

(continued on next page)

 COPYRIGHT © GLENCOE/McGRAW-HILL

Procedures. An SOP should be clearly worded so that it is readily understandable by any reader with knowledge of the general concept of the procedure. In writing, follow these general guidelines:

❑ Use the active voice and the present tense.

❑ Do not use the pronoun "you."

❑ Describe the process. Include any appropriate safety information, regulatory information, or standards.

❑ Define unfamiliar terms. Define acronyms, abbreviations, and specialized or unusual terms in a separate definition section or in the appropriate section.

❑ Present sequential procedures. Include the following, as appropriate:

■ Scope and applicability of procedure, describing the purpose of the procedure and when it should be done.

■ Summary of the procedure. If necessary, use diagrams and flow charts to summarize a series of steps. Present the steps of the procedure in order. If appropriate, number the steps.

■ References to checklists. Checklists ensure that the steps in a procedure are performed in order. They also document completed actions. A checklist included as part of an activity should be referenced at the points in the procedure where it is to be used. It should be attached to the SOP as a part of the SOP.

■ Definitions of acronyms, abbreviations, and specialized terms.

■ Health and safety warnings, identifying operations that could result in personal injury or loss of life and explaining what will happen if the procedure is not followed or is followed incorrectly. List such warnings here and at critical steps in the procedure.

■ Comprehensive safety standards for unique tools and machines.

■ Cautions, indicating activities that could result in equipment damage. List these cautions here and at critical steps in the procedure.

■ Interferences, describing any component of the process that may interfere with the accuracy of the final product.

■ Personnel qualifications, specifying the minimal experience the SOP user needs to complete the task satisfactorily and citing any applicable requirements, such as certification or training.

■ Equipment and supplies, listing and specifying equipment and materials.

■ Procedure, listing all pertinent steps in order along with the materials needed to complete them. The detail should be sufficient to allow someone with limited experience with or knowledge of the procedure, but with a basic understanding, to successfully perform the procedure when unsupervised.

■ Troubleshooting, listing actions to identify and solve problems.

■ Data and records management, identifying forms to be used, reports to be written, and data and record storage information.

Quality Control. Include the following:

❑ List appropriate quality control (QC) procedures and QC material required to demonstrate successful performance of the method. Include specific criteria for each.

❑ State the frequency of required calibration and QC checks and discuss the rationale for quality decisions.

❑ State the criteria for QC results.

❑ State the actions required when QC data exceed QC limits or appear in the warning zone.

❑ Identify the procedures for reporting QC data and results.

(continued on next page)

References. All other materials that pertain to the SOP should be added here.

❑ Include references to documents and procedures that interface with the SOP, such as related SOPs, published literature, and methods manuals.

❑ Attach any checklists mentioned in the SOP. Checklists should be attached to the SOP as a part of the SOP. The checklists are not the SOP.

❑ Attach any other appropriate information.

Finalization & Document Control

Review and Approval. SOPs should be reviewed by individuals with appropriate training and experience with the process.

❑ Before being finalized, the draft SOP should be tested by someone other than the original writer.

❑ The finalized SOP should be approved and signed by the appropriate authority. Signature approval of an SOP indicates its review and approval by management.

❑ The finalized SOP should be forwarded to the proper parties.

Review and Revision. The quality management plan should specify the individual responsible for ensuring that SOPs remain current. SOPs should be reviewed on a regular basis. This will help ensure that they are appropriate and current. Review can also determine whether the SOP is even needed. When procedures governed by an SOP are changed, these guidelines should be followed:

❑ Revise the SOP and submit it for reapproval.

❑ Add the review date to each SOP that has been reviewed.

❑ Withdraw from the file any SOP that describes a process that is no longer followed. Archive those SOPs.

Document Control. The document control process for numbering, tracking, and archiving SOPs should be described in the quality management plan. The following are general guidelines:

❑ Make sure that the SOP carries the revision number and date.

❑ Number the pages of the SOP.

❑ Make sure that current SOPs are readily accessible (in hard copy or electronic format) in the work areas of those performing the activity.

❑ Make sure that the quality assurance manager maintains a master list of all SOPs with the date of the current version.

❑ Check the quality assurance plan for information on where and how outdated SOP versions are to be archived.

❑ Check that archived SOPs are available for review. SOPs in electronic format are usually easier to access than those stored as hard-copy documents.

❑ To protect against unauthorized changes, limit electronic access to SOPs to a read-only format.

MEETS THE FOLLOWING (MSSC) PRODUCTION STANDARDS

P1-K4E	Product and process documentation is completed, maintained, and forwarded to the proper parties.
P2-A14C	Write comprehensive safety standards for unique tools and machines.
P7-A14D	Write safety rules, safe procedures and practices, etc.
P8-O1J	Skill in developing and documenting quality procedures, checklists and methods.
P8-O6D	Skill in communicating clearly to large production groups about aspects of the quality system, including documentation, specification, or design changes.
P8-A14C	Write quality procedures.

 COPYRIGHT © GLENCOE/MCGRAW-HILL

Calibrating Inspection, Measuring & Test Equipment

Calibration is the adjustment of inspection, measuring, and test equipment (IM&TE) to ensure that it measures accurately against a certain standard. Such equipment includes micrometers, recorders, signal generators, counters, oscilloscopes, balances, scales, transducers, gauges, calipers, wrenches, meters, sensors, and other data collection equipment. These instruments are designed to measure mass, dimension, torque, force, pressure, vacuum, voltage, resistance, current, flow, waveforms, temperature, and humidity. Measurement records ensure that products meet the manufacturer's or user's specifications. Calibration increases confidence that IM&TE measurements are accurate, reliable, and reproducible.

Management of IM&TE, including calibration and use, is determined by company policy. This checklist identifies general areas of responsibility for IM&TE managers, users, and calibrators.

IM&TE Manager Responsibilities

❑ Identify IM&TE requiring calibration. This includes all IM&TE:

- Used to perform measurements where the quality or accuracy of a process is critical.
- Used in applications, such as testing air quality, designed to ensure the safety of personnel.
- Used to demonstrate the conformance of the product to specified requirements.
- Used to perform measurements associated with acceptance testing of new instrumentation.
- Used for inspection, maintenance, calibration, and/or qualification of aviation hardware.
- Used in measurement of processes where test equipment accuracy is essential for the safety of personnel.
- Used in telecommunication, transmission, and test equipment where exact signal interfaces and circuit confirmations are essential.
- Used in development, testing, and special applications where the specifications, end products, or data are accuracy sensitive.

❑ Ensure that documented procedures are established and maintained.

❑ Ensure that IM&TE is calibrated in accordance with these procedures.

❑ Determine the frequency and extent of testing required to ensure that the IM&TE is capable of verifying the acceptability of the product.

❑ Ensure that IM&TE that needs to be calibrated in accordance with this procedure is identified. This can be done through work instructions.

❑ Identify IM&TE that does not require calibration with a "Calibration Not Required" sticker.

❑ Keep records of the testing.

IM&TE User Responsibilities

❑ Determine the accuracy needed for measurements.

❑ Use only IM&TE that is capable of meeting the accuracy requirements.

❑ Determine the calibration interval.

❑ Determine the acceptable tolerances for the IM&TE and give this information to the calibration lab.

❑ Submit IM&TE for calibration before initial use and at the required calibration intervals.

❑ Use only IM&TE calibrated in accordance with this procedure.

❑ Reassess calibration and measurement needs when processes or product requirements change.

❑ Handle and store IM&TE in such a way that accuracy and fitness for use are maintained.

❑ Ensure that environmental conditions are suitable for the use of IM&TE.

❑ Safeguard IM&TE against adjustments that would invalidate calibration.

(continued on next page)

Calibrator Responsibilities

❑ Identify calibration procedures.

❑ Ensure that environmental conditions are suitable for the calibrations, inspections, measurements, and tests carried out.

❑ Ensure traceability to an international or national standard.

❑ Safeguard calibration facilities, including IM&TE, against adjustments that would invalidate the calibration.

❑ Provide the IM&TE user with documentation of the calibration status. This should include:

- The date of calibration (for IM&TE calibrated before each use) or the date due for calibration.

- The calibration interval.

- Information on equipment that could not be calibrated or was out-of-tolerance.

❑ Place a calibration sticker on the IM&TE that shows the date calibrated or the date due for the next calibration.

❑ Document calibration procedures.

❑ Notify the IM&TE user when calibrations are due.

Responding to Out-of-Tolerance IM&TE

1. The calibration lab shall notify the IM&TE user of IM&TE found to be out-of-tolerance during calibration.

2. When notified of an out-of-tolerance IM&TE, the IM&TE user shall:

   ❑ Analyze and document the impact on product quality of the out-of-tolerance IM&TE.

3. If the out-of-tolerance IM&TE could have impacted product quality, the IM&TE user shall:

   ❑ Take appropriate action based on the significance of the error and the instrument's application.

   ❑ Adjust the calibration interval if necessary and request the change from the calibration lab.

MEETS THE FOLLOWING (MSSC) PRODUCTION STANDARDS

P1-K5A	The calibration of the testing equipment is verified.
P1-O1A	Skill in using inspection equipment, including how to calibrate, what type of equipment to use, and what frequency to use.
P4-K2	Check calibration of gauges and other data collection equipment.
P4-K2A	Calibration schedule is implemented according to specifications.
P4-K2C	Instruments that are out of calibration are immediately recalibrated or referred to the appropriate parties for recalibration or repairs.
P4-O4F	Skill in maintaining and storing inspection tools.
P4-O5A	Knowledge of the calibration standards, requirements, and equipment.
P4-O5B	Knowledge of environmental impact that affects calibration requirements.
P4-A2E	Check all inspection equipment.
P4-A4C	Decide if calibration is out-of-date and when recalibration is required.
P4-A5A	Organize and maintain measuring equipment calibrations.
P8-O2J	Knowledge of calibration plan and procedures using current references and standards.

COPYRIGHT © GLENCOE/McGRAW-HILL

Measuring for Precision & Accuracy

In ordinary speech, the terms *accuracy* and *precision* are often used to mean the same thing. In reference to measurements as part of an inspection, however, the terms have very different meanings. When you measure something, there is always a certain amount of error. This error is due to limits of the measuring tool, calibration or adjustment of the measuring tool or instrument, and training in correct use and reading. In this sense, error is built into the system. Error does not refer to mistakes, which can be traced to a specific problem, such as misreading the value or writing down the wrong number. Accuracy and precision are two ways of determining the significance of the error and whether the results are useful.

Precision

Precision is based on the ability of an instrument to measure a value and is indicated by the reproducibility of a result. Every measuring tool has a smallest increment that it is capable of measuring. Precision is related to that increment. For example, if you use a ruler that has markings at 1 mm intervals, you may be able to determine the length of a part to within a value of 0.3 mm by estimating the distance between the markings. The precision of the ruler is expressed as ±0.3 mm. There is no way that you can use this ruler to measure an object with more precision than that. However, a micrometer can measure smaller intervals, perhaps ±0.01 mm.

Accuracy

Accuracy is an indication of how close a value is to a standard. Accuracy is measured by comparing data taken with a measuring tool to a known value. For example, a large scale may be tested with a 100 kg weight to determine that it is accurate. If the scale reads 95 kg, the scale is not accurate. It may need to be recalibrated or checked for a mechanical problem.

> **Precision** is related to the quality of the method.
> **Accuracy** is related to the quality of the result.

Comparing Precision & Accuracy

To be useful, a measurement must be both precise and accurate. An erroneous measurement can be precise but not accurate, accurate but not precise, or both inaccurate and imprecise. In order to improve the value of the measurement, you must determine the type of error.

If a measurement is done with a high-quality tool, it will give precise results. Repeated measurements will read the same value with very little variation. However, if the instrument is not calibrated correctly, or if there is some other systematic error, the results may be different from the actual dimension. In this case, the results are precise but not accurate. For example, if repeated weighing of the 100 kg weight gives values of 96.2 kg, 96.1 kg, and 96.2 kg, the values are precise but not accurate. Correcting the error may require recalibration, replacement of a part, or a mechanical adjustment.

If the scale gives three readings of 98.2 kg, 100.1 kg, and 102.0 kg, the average value—100.0 kg—is accurate, but it is not precise. The scale could not be used to weigh a product that has a specification of 100.0 ±0.5 kg, because you cannot be confident that the weight of product is within the specification, even if the reading indicates that it is.

(continued on next page)

Inaccurate readings cannot be corrected by calibration. The source of the imprecision may be a need for maintenance or repair, errors in operation, or simply that the measuring tool is not the correct choice for the application.

Inaccuracy can also be the result of improper procedures or inadequate training. If three workers weigh the same object and obtain reproducible results that differ from one another, the problem may be solved by training in the correct procedures to use the scale. The scale itself is probably precise. The reproducibility of weighings by each operator would indicate that the problem is accuracy rather than precision.

Using the Right Measuring Tool

Measuring tools must be matched to the precision of the measurement required by a specification. If a specified dimension is 25.0 mm ±0.01 mm, the ruler with a ±0.3 mm precision will not provide usable data. A measurement that reads 25.0 mm could indicate a value anywhere within the range from 24.5 mm to 25.5 mm. Therefore, the part could be out of spec even though the measurement does not indicate that it is. On the other hand, it is not necessary to use a micrometer, which is more expensive and takes more time to use, to measure a part whose specification is 25 mm ±2 mm.

APPLYING YOUR KNOWLEDGE

Measurement	Readings, Worker 1	Readings, Worker 2	Readings, Worker 3	Standard value
Temperature (°C)	56.3, 56.2, 56.3	56.3, 56.4, 56.2	56.2, 56.3, 56.4	56.3 ± 0.3
Weight (kg)	32.3, 32.3, 32.3	32.2, 32.4, 32.3	32.4, 32.4, 32.2	32.2 ± 0.5
Flow rate (L/min)	48.2, 48.4, 48.4	51.8, 51.9, 51.8	50.3, 50.4, 50.4	51.8 ± 0.5
Length (cm)	3.65, 3.68, 3.66	3.69, 3.67, 3.62	3.67, 3.66, 3.62	3.66 ± 0.2

Use the results in the table to answer the following questions.

1. Which measurements are accurate but not precise?

2. Which measurements are precise but not accurate?

3. Which measurements are both precise and accurate?

PRODUCTION CHALLENGE

As a cutting machine setter and setup operator for a paper company, you are responsible for measuring the paper you've cut to verify conformance to specifications. What measuring tool would you use? If the measurement readings for the last lot of paper cut are not accurate, what is the first step you would take to determine the reason for the error?

MEETS THE FOLLOWING MSSC PRODUCTION STANDARDS

P8-O2E	Skill in determining accuracy and precision when using measuring equipment.
P8-A11C	Update measurement skills to improve quality.

COPYRIGHT © GLENCOE/MCGRAW-HILL

Conducting Dimensional Inspections

The materials used to produce a product should be inspected at all stages of the process to determine their quality and condition. When production begins, the first piece of a production run or of a shift should be visually and dimensionally inspected against the blueprint specifications. This inspection is known as a first-article, or first-piece, inspection.

The quality of a product with specified dimensions can be checked through a dimensional inspection. Such parts include those that have been machined, formed, or cast. This inspection compares the inspection measurements with the customer specifications on the product blueprint. Adjustments to the manufacturing process can then be made to correct quality problems.

A deviation is a variation in the actual measured value of the part from the dimension specified on the blueprint. Any allowable deviations regarding specified values will be noted on the blueprint. The following values are usually checked.

Machined Surfaces

❑ **Actual size**—the true exact measured width or length of an item. A micrometer can be used to measure short lengths.

❑ **Taper**—the allowable deviation in thickness, diameter, or width from one end to the other of an elongated object. This can be checked with a test indicator or, if greater precision is needed, an indicator.

❑ **Lengths and/or depths**—the true exact measured length or depth of an item. A micrometer can be used to measure short lengths. Depth gauges can be used to measure depth.

❑ **Surface finish**—a measurement of the smoothness of a surface.

   ■ Depending on the need for precision, this inspection might be visual. An optical comparator can be used for detailed visual inspection and measurement. It projects a magnified silhouette of the part onto a screen.

   ■ For greater accuracy, a surface profilometer might be used. This electronic instrument measures the roughness of a surface.

❑ **Surface irregularities:**

   ■ **Burrs**—thin ridges or areas of roughness.

   ■ **Chatter**—surface imperfections usually produced by the part and the tool bouncing against one another because of improper tool setup or part setup.

   ■ **Fixture marks**—marks left by the fixture holding the workpiece.

Inside Diameter

This is the measurement of an object's diameter across its inner perimeter.

❑ **Actual size**—the true exact measured inside diameter of an item. It can be measured using internal calipers, slot gauges, indicating bore gauges, and 3-point bore gauges.

❑ **Taper**—the allowable deviation in thickness, diameter, or width from one end to the other of an elongated object. It can be measured using taper gauges or more sophisticated air gauging tools.

❑ **Out-of-roundness**—the allowable deviation from true roundness. It can be measured using specially designed NOGO gauges.

❑ **Length of turn/bore.**

Outside Diameter

This is the measurement of an object's diameter across its outer perimeter.

❑ **Actual size**—the true exact measured outside diameter of an item. A micrometer can be used to measure small items.

❑ **Taper**—the allowable deviation in thickness, diameter, or width from one end to the other of an elongated object. Taper can be checked with a test indicator or, if greater precision is needed, an indicator.

(continued on next page)

- **Out-of-roundness**—the allowable deviation from true roundness. Tolerances within 0.01" can be checked with a digital or vernier caliper-style gauge or with an inside micrometer.
- **Length of turn/bore.**

Geometric Characteristics

- **Parallelism**—an allowable value that a surface, plane, or axis be equidistant to a given surface, plane, or axis. A pair of micrometers can be used to check for parallelism in small parts.
- **Perpendicularity**—an allowable value indicating the amount of variation that a surface, plane, or axis could vary from a true right angle to a given surface, plane, or axis. A basic machinist square can be used to check perpendicularity if the tolerance is 0.001" or less. For more precise measurements, use electronic height gauges.
- **Flatness**—a specified value indicating the amount of variation from a true plane. It can be checked using a height gauge and a surface plate.
- **Concentricity**—an allowable amount of variation indicating that two or more circular features share the same centerline. Check using concentricity checking gauges.

Threads

- **Size (pitch diameter)**—the diameter of a bolt measured at the pitch line. The pitch line intersects the threads at a distance from the axis of the bolt such that the intersections with the sloped faces are one-half the pitch apart. This is measured using thread ring gauges or a thread comparator.
- **Length** of effective thread.

Corrective Actions & Documentation

To ensure quality, the manufacturing process must be in compliance with standards. Products that do not meet specs should be promptly identified. The process may need to be stopped. The corrective action needed to bring the manufacturing process back into control must be implemented. The person responsible for performing the dimensional inspection may be able to suggest or perform the corrective actions needed to correct quality problems.

Inspection results and quality tests should be documented properly. The documentation should be completed accurately and forwarded to the appropriate parties. The inspection documents should then be maintained according to company policy.

MEETS THE FOLLOWING MSSC PRODUCTION STANDARDS

P1-K3C	First piece of production run meets specifications.
P1-K4D	Product meets customer specifications.
P1-K5	Inspect the product to make sure it meets specifications.
P1-K5C	Product and production processes that do not meet specifications are identified promptly.
P1-K5D	Inspection documentation is completed accurately and forwarded to the correct parties.
P1-K5E	Appropriate testing and inspection tools and procedures are followed.
P1-O1B	Skill in using multi-gauging to inspect, verify, and document whether product dimensions meet customer requirements.
P1-A15C	Read and understand quality documentation and production spec sheets.
P4-K4	Inspect materials at all states of process to determine quality or condition.
P4-O1I	Skill in interpreting test results.
P4-A4A	Determine when production must be stopped if it isn't meeting specifications.
P4-A4B	Identify the corrective action necessary to bring a process back into control.
P8-K2	Suggest or perform corrective actions to correct quality problems.

COPYRIGHT © GLENCOE/MCGRAW-HILL

Conducting Destructive Testing

To ensure the safety of a product, it is necessary to test it before it leaves the production facility. Testing will often depend on the number of units produced, the time needed for testing, and the type of testing. Destructive testing is a type of testing that cannot be performed on each production unit. Such testing would destroy all of the products.

Destructive testing is performed on performance-critical products manufactured to exacting physical specifications. It is also performed when it is desirable to know the physical limitations and tolerances of a given product. In either case, product units are stressed to their breaking point. Data regarding the amount of stress at the time of critical failure is recorded.

Types of Stress

Four common types of stress are measured. These stresses are tensile, compressive, torque, and shear.

- Tensile stress is the amount of force applied to an object by pulling or stretching the object.

- Compressive stress is the amount of force applied to an object by pressing or crushing the object.

- Torque stress is the amount of force applied to an object by twisting or rotating it against a fixed point.

- Shear is the amount of force applied to an object by parallel, but not overlapping, forces acting in opposition to one another on opposite sides of an object (think of scissors shearing paper).

Different materials tolerate different amounts of different types of stress. Some products are manufactured to tolerate more stress of one type than another. To determine if a finished product meets specifications, stress testing is needed. Such testing is often destructive, so it is kept to the absolute minimum needed to ensure a safe product.

Stress & Force

A manufacturer makes stainless steel bolts used to fasten components at critical points. Because the failure of a bolt could lead to injury or death, the manufacturer always tests a small sample of each batch of bolts they produce. The bolts are tested for performance under four of the stresses that they will encounter under normal use: tensile, compressive, torque, and shear.

Tensile Strength. Tensile strength is the measure of the resistance to deformation or breakage under tensile stress and is usually given in pounds per square inch (psi, lb/in^2). To test its tensile strength, the bolt is attached at opposite ends to a fixture containing a hydraulic piston. When the piston is activated, it forces the fixture apart. This puts tensile stress on the bolt mounted inside.

Compressive Strength. Compressive strength is the measure of the resistance to deformation or breakage under compressive stress and is usually given in pounds per square inch (psi, lb/in^2). In a test for compressive strength, the bolt is attached at opposite ends to a fixture containing a hydraulic piston. When the piston is activated, it forces the fixture together. This puts compressive stress on the bolt mounted inside.

Torque Strength. Torque strength is the measure of the resistance to deformation or breakage under torque stress and is usually given in foot-pounds (ft-lb). In a test for torque strength, the bolt is mounted securely in a fixture and a torquing device is attached to the head of the bolt. The bolt is then twisted until it is deformed or broken.

(continued on next page)

Shear Strength. Shear strength is the measure of the resistance to deformation or breakage under compressive stressors acting in parallel, but not overlapping, on opposite sides of an object. It is usually given in pounds (lb). In a test for shear strength, the bolt is mounted halfway in a fixture, and a force is applied perpendicular to the axis of the bolt.

Destructive Testing Experiment

To see how this process works, perform the following experiment to test torque strength.

Tools & Materials

- Safety glasses
- Torque wrench with a dial indicator in ft-lb
- 3 hex head bolts (3/8" diameter by 1". One bolt should be Grade 1 or 2, one Grade 5, and one Grade 8)
- 3 nuts (1 for each bolt)
- Flat washers for each bolt (a number sufficient to allow space for attaching wrenches)
- Box wrench

Procedure

1. Wear proper eye protection.
2. Locate a steel plate that is at least 1/8" thick, and drill a hole in the plate large enough to accept a 3/8"-diameter bolt.
3. Mount the plate in the vice, leaving enough room to attach the wrenches to the bolt.
4. Place a bolt in the hole and add enough washers to allow clearance for the wrenches. Finger-tighten the nut onto the bolt.
5. Reset the torque on the torque wrench to 0.
6. Hold the nut in place with the box wrench and place the torque wrench on the head of the bolt.
7. Gradually increase the amount of torque until the bolt breaks or deforms.
8. Record the amount of torque applied to the bolt at the moment of critical failure.
9. Reset the torque wrench and follow the same procedure with the other two bolts to be tested.

MEETS THE FOLLOWING (MSSC) PRODUCTION STANDARDS

P2-A16D	Measure torque specifications and spec tolerances to properly maintain and use equipment.
P8-A17A	Knowledge of physical science to conduct stress force experiments.

    COPYRIGHT © GLENCOE/MCGRAW-HILL

Creating & Maintaining a Calibration Log

The National Institute of Standards and Technology (NIST) is an agency of the U.S. Department of Commerce. It provides certification and calibration for precision instruments. When purchasing a precision instrument, check to see that it has a NIST label. This label indicates that the instrument has been certified by NIST. A NIST-certified instrument will maintain accuracy within specified limits for at least one year. To ensure accuracy, the instrument must be recertified. For this, the accuracy of the instrument needs to be checked. This can be done by the manufacturer of the instrument or by a NIST calibration laboratory.

Purpose of a Calibration Log

Calibration is the process of standardizing an instrument to ensure that it will measure within a specific range. Traceability is an important issue in calibration. Traceability requires the establishment of an unbroken chain of events to stated references. The instrument calibration record provides proof of traceability.

Each instrument requiring calibration should be identified by a permanent identification number. This will simplify record keeping and allow the accuracy of each instrument to be easily tracked.

Maintaining an instrument calibration log is essential. A calibration log is a detailed record. In a calibration log, each type of instrument (e.g., micrometer) typically has its own log sheet. In the past, this was a paper record. Today, the log sheet is usually maintained on a computer. This simplifies the tracking of calibration schedules and the results of quality checks. The following information must be entered by the individuals conducting the calibration.

❏ Date.
❏ Time.
❏ Instrument identification number.
❏ Instrument use.
❏ Reference reading.
❏ Difference from reference reading.
❏ Initials of person testing equipment.
❏ Comments.
❏ Initials of person verifying correction.
❏ Date of verification.

Some instruments may require replacement. When replacing an instrument, follow these guidelines as they relate to the instrument calibration log.

❏ Replace a measuring instrument when it consistently fails to provide accurate readings during calibration. For example, a thermometer should be replaced when the difference between the reference reading and the reading on the thermometer is consistently greater than $\pm2°F$ [$\pm0.5°C$].

❏ When a measuring instrument is replaced, note this in the Comments column of the instrument calibration log.

❏ Attach a new identification label to the replacement instrument.

❏ Assign a new number to the replacement instrument.

See the instrument calibration log on p. 222.

(continued on next page)

Instrument Calibration Log

Company Name

Log Page No.

Date	Time	Instrument Identification Number	Instrument Use	Reference Reading	Difference from Reference Reading	Initials	Comments	Verified by	Date of Verification

MEETS THE FOLLOWING (MSSC) PRODUCTION STANDARDS

P4-K2B	Instrument certification is checked both by reviewing documentation and through careful observation during use.
P4-K2C	Instruments that are out of calibration are immediately recalibrated or referred to the appropriate parties for recalibration or repairs.
P4-O4F	Skill in maintaining and storing inspection tools.
P4-A1A	Use computer system to track gauge calibration schedules and results of quality checks.
P4-A2E	Check all inspection equipment.
P4-A4C	Decide if calibration is out-of-date and when recalibration is required.
P4-A5A	Organize and maintain measuring equipment calibrations.
P4-A14D	Create detailed log of calibration of gauges and other data collection equipment.
P8-O2J	Knowledge of calibration plan and procedures using current references and standards.

COPYRIGHT © GLENCOE/McGRAW-HILL

Completing a First-Article Inspection Report

A first-article inspection report (FAIR) is a document that certifies that each first-article unit delivered to the buyer was produced and inspected in accordance with the buyer's specifications. It demonstrates conformance to the purchase order. The report includes all physical, material, and chemical test data associated with the part.

Without first-article inspection, the risk of producing an entire batch of flawed parts is extremely high. The production of scrap parts cuts into profit. A full first-article inspection can be time consuming. The challenge is to implement a quality control program at all critical stages that does not slow down production.

The circumstances requiring the completion of a FAIR differ among companies. Generally, a FAIR is required when:

- The product is being manufactured for the first time.
- There has been a lapse of twelve months or more in production of the product.
- There has been a significant change in the manufacturing method.
- There has been a change in manufacturing or processing equipment.
- The manufacturing facilities have been moved to a different location.
- There has been a change in inventory sources.
- There has been an occurrence that might adversely affect the manufacturing process.

Use this checklist to make sure that the FAIR is handled properly.

- ☐ The FAIR must be completed accurately by the manufacturer's qualified personnel.
- ☐ They must use the measurement system used in the blueprint and address all blueprint notes.
- ☐ They must be able to interpret all spec charts to understand their impact on the final product.
- ☐ The detection of a nonconformance during the first-article inspection will require corrective action. A partial (delta) FAIR is required for the corrected nonconforming condition.
- ☐ The person making the inspection must record all measurements and observations in the FAIR.
- ☐ That person must sign and date the report. The signature certifies that any needed corrective actions have been taken.
- ☐ The FAIR must be forwarded to the appropriate parties, including customers, in a timely manner.
- ☐ FAIRs should be maintained by the quality assurance department according to company policy. These reports are usually kept on file by the supplier for a minimum of seven years or as required by the buyer.

The format of a FAIR varies. The report shown here presents a general format.

XYZ Manufacturing Co., Inc. • Any Town, IL 61606
First-Article Inspection Report

Date	Part Number	Part Name	Purchase Order	Item Number

Material Type and Temper		Material Heat Lot Number	

Hardness	Required	Actual	Conductivity	Required	Actual

Reference	Characteristic and Tolerance	Actual Dimension	Accept	Reject	Print Location Zone	Inspection Device	Remarks

Quality Assurance Inspector	Print Name		Signature			Date	

(continued on next page)

MEETS THE FOLLOWING (MSSC) PRODUCTION STANDARDS

P1-K3C	First piece of production run meets specifications.
P1-K4D	Product meets customer specifications.
P1-K5	Inspect the product to make sure it meets specifications.
P1-K5B	Established sampling plan and inspection policies and procedures are followed.
P1-K5C	Product and production processes that do not meet specifications are identified promptly.
P1-K5D	Inspection documentation is completed accurately and forwarded to the correct parties.
P1-K5F	Adjustments needed to bring the production process back into specification are identified and communicated.
P1-A14B	Document inspection results.
P1-A15C	Read and understand quality documentation and production spec sheets.
P4-K4B	Inspection tools and procedures are selected and used correctly.
P4-K5	Document the results of quality tests.
P4-K5A	Data forms are checked to ensure that they are complete and accurate.
P4-K5B	Information is evaluated and interpreted correctly.
P4-K5C	Data is forwarded to correct parties.
P4-O4D	Skill in evaluating the characteristics of a finished product against specifications.
P4-A2D	Interpret all spec charts to understand their impact on the final product.
P4-A4B	Identify the corrective action necessary to bring a process back into control.
P4-A4D	Determine when and where to inspect or audit process for quality of product or process to meet customer requirements.
P4-A4E	Decide if a product is within tolerances.
P4-A5D	Schedule inspection of production at all critical stages.
P4-A16C	Determine if a part is acceptable based on actual vs. dimension/tolerance specifications.
P5-K3B	Quality issues are raised in a timely way.
P5-K3C	Quality issues are addressed in a timely way.
P5-K3E	Communication is clear and relevant to quality.
P5-K3G	Quality issues are recorded, and tracked and reported back to original communicator.
P5-K4A	Communication reflects knowledge of production requirements, levels, and product specifications.
P5-K4E	Communication is clear and relevant to production and products.
P5-A14C	Document into quality system the defects in parts produced.
P5-A16B	Perform measurements to verify parts meet customer requirements.
P8-K1	Communicate quality problems.
P8-K1B	Quality problems are communicated promptly to appropriate parties.
P8-K1C	Quality problems are documented according to established processes.
P8-K2	Suggest or perform corrective actions to correct quality problems.
P8-K2B	Quality issues or adjustments are documented properly.
P8-K2F	Product quality is documented following corrective action.
P8-K4	Record process outcomes and trends.
P8-K4A	Records on quality process are maintained to appropriate standards.

COPYRIGHT © GLENCOE/MCGRAW-HILL

Selecting Product Shipping Containers

Many factors must be considered when choosing the proper container for shipping a product. Products are shipped individually and in bulk. They are shipped in boxes and in special containers. They are shipped by land, air, and sea. Regardless of the way in which products are shipped, answering the following questions will help you choose the appropriate product shipping container. Every manufacturer has standard procedures regarding product shipping which present guidelines that are detailed and specific. The guidelines below are generalized.

❑ Does the product's packaging warrant using a shipping container?

❑ What are the dimensions of the product to be shipped?

❑ What is the estimated gross weight of the product and its required packing material?

❑ Will the product be shipped individually or as part of a pallet?

❑ Is the container appropriate for the mode of transport that will be used (e.g., aircraft, truck, cargo ship)?

❑ Is the container appropriate for the distance the product is being shipped?

❑ Does the product require protection beyond that provided by the container? (Does it need extra packing material or more than one container?)

❑ If the product requires extra protection, can it be over-boxed? This consists of adding an extra box to the outside of the product's existing packaging.

❑ If the product cannot be over-boxed, can it be surrounded by filler?

❑ If the product can be surrounded by filler, does the container allow enough room for the filler? This area is called the barrier dimension. Three inches of space all around is a good rule of thumb. If the product is fragile, leave 1 to 3 extra inches to accommodate more shock-absorbing material.

❑ If the product is live, does the container allow sufficient airflow?

❑ Check with the shipping company for regulations regarding the shipment of live cargo.

❑ Ensure that no prohibited live cargo is being shipped.

Most cardboard containers carry a label that gives such information as crush strength, bursting strength, maximum weight limit, interior dimensions, and other information pertinent to shipping products. When shipping products in cardboard boxes, check the manufacturer's specifications to ensure that the box you have chosen meets the following requirements.

❑ Ship products in new boxes whenever possible. Used, worn boxes may not retain their full protective strength.

❑ The gross weight of the product and the packing material should not exceed the maximum allowable weight of the shipping container.

❑ If the product is to be stacked or palletized, make sure that the container has sufficient crush strength to withstand the weight of other packages.

❑ If the product is to be shipped to or through a moist or humid environment, is the box sufficiently waterproofed to prevent failure of the box in adverse conditions?

❑ If the product were dropped during shipping, is the bursting strength of the box sufficient to protect its contents from any intrusion?

(continued on next page)

❑ Does the product require a special container? Corrugated cardboard boxes are usually appropriate and work well for many products, but some products may require special containers. Special containers are required for products such as the following:

- Temperature-sensitive products may need insulated containers.

- Products being shipped to or through a humid environment may need a plastic container to protect the product from moisture.

- Hazardous materials require special packaging.

❑ Consider whether the box will protect the outside environment and those handling the package from its contents. This is especially important in the case of hazardous materials and sharp objects.

❑ Check with the company that will handle the actual transport of the product for any additional requirements.

MEETS THE FOLLOWING **PRODUCTION STANDARDS**

P1-06H	Knowledge of available packing materials to determine the safest method of shipping the product.
P1-A4E	Determine proper packaging for product to limit product damage during shipment.

COPYRIGHT © GLENCOE/MCGRAW-HILL

Selecting & Using Filler Materials

Filler material is packing material used to protect a product during shipment. The choice of filler material will depend on several factors. These include the nature of the product, the mode of shipment, the destination, and the environmental conditions under which the product is shipped. The filler material used should:

- Dampen the effects of shock and vibration on the product.
- Protect the product reasonably well from the handling that normally occurs during shipping.
- Protect the product from heat, cold, moisture, and changes in altitude.
- Protect the environment from the product in the case of hazardous materials. It should also protect the product from the environment.

Every manufacturer has standard shipping policies regarding the selection and use of filler materials. These are detailed and specific. The following guidelines on selecting and using the appropriate filler material are general guidelines. The answers to the following questions will help determine the filler material to select from the chart.

- ❏ Are hazardous materials part of the product?
- ❏ Is the product already packaged?
- ❏ Will the product be shipped individually or in bulk?
- ❏ How heavy is the product being shipped?
- ❏ Is the product fragile, semi-fragile, or non-fragile?
- ❏ If the product is fragile, is the product particularly fragile (e.g., an egg or a thin pane of glass)?
- ❏ Does the product have a surface area that is roughly even on all sides or are one or two dimensions significantly larger or smaller than the others?
- ❏ Will the product be shipped to or through areas with extreme changes in temperature or altitude?
- ❏ Does the product need to be kept at a specific temperature?
- ❏ Is special protection from the environment needed to maintain the quality of the product?

In the following chart, the barrier dimension refers to the space around the product used for filler.

Filler Materials and Their Uses		
Filler Material	**Barrier Dimension**	**Method**
Loose-fill Peanuts	3 inches	Fill dead space between product and container wall with peanuts. Over-fill to allow for settling. Do not use with narrow objects that might easily shift during shipping.
Encapsulated-air Plastic Sheeting	2 inches + 2 inches	Wrap product in 2 inches of sheeting. Leave at least 2 inches of dead space between sheeting and container wall. Fill dead space with filler material. Works well with lightweight items.
Polyethylene Foam Sheeting	2 inches + 2 inches	Wrap product in 2 inches of sheeting. Leave at least 2 inches of dead space between sheeting and container wall. Fill dead space with filler material. Works well with lightweight items and provides good surface protection.

(continued on next page)

Filler Materials and Their Uses (continued)		
Filler Material	**Barrier Dimension**	**Method**
Inflatable Packaging ("air pillows")	3 inches	Cushion product by filling dead space with inflatable packaging. Extreme changes in temperature and altitude will adversely affect the performance of this material.
Foam-in-Place	3 inches	Apply foam to product or container walls. Film expands to fill dead space in container. Good for odd-shaped products. For maximum effectiveness, distribute foam evenly. Foam densities vary. Read manufacturer specifications to determine which foam is right for the application.
Kraft Paper	4 inches	Fill dead space in container with crumpled kraft paper. Paper should be tightly wadded. Good for light to medium-weight products that are semi-fragile to non-fragile.
Paper Cushioning	4 inches	Wrap paper around a product as a moisture barrier and crumple it to fill dead space. Good for medium-weight to heavy products that are semi-fragile to non-fragile.
Expanded Polystyrene Foam	3 inches	Used for packaging heavy products that are semi-fragile to non-fragile.
Polyethylene Foam	3 inches	Fill dead space with foam. Good for lightweight products. Low-density foam with good shock absorption.
Polyurethane Foam	3 inches	Fill dead space with foam. Better for lightweight products. Good shock absorption.
Corrugated Cardboard	As needed	Cushion semi-fragile to non-fragile products and bolster the shipping container.

MEETS THE FOLLOWING (MSSC) PRODUCTION STANDARDS

P1-O6H	Knowledge of available packing materials to determine the safest method of shipping the product.
P1-A4E	Determine proper packaging for product to limit product damage during shipment.

COPYRIGHT © GLENCOE/MCGRAW-HILL

Lifting, Moving & Material Handling

When you are packaging products or preparing them to be shipped, keep in mind the guidelines for safe lifting, moving, and handling of materials. This will protect you from injury and the product from damage.

Personal Safety

❑ Wear the recommended personal protective equipment (PPE) when handling materials that present health hazards, such as acids, corrosives, and irritants.

❑ Before handling unfamiliar and hazardous materials or chemicals, read the label for safety instructions. Refer to the Material Safety Data Sheet (MSDS) or consult your supervisor.

❑ Inspect items to be handled for slivers, jagged edges, burrs, and slippery surfaces.

❑ Wipe off oily, wet, or dirty items before handling them.

Travel

❑ Plan your route of travel to be sure it is clear of obstacles that would hamper the safe movement of materials.

❑ Make sure the floor is clean, even, flat, dry, and free of slippery areas.

Lifting

❑ Test the weight of objects to be lifted. Get help if an item is too heavy to lift alone.

❑ Work with someone else to push a heavy load; one should steer and pull while the other pushes.

❑ When a team is handling an item, one person should give voice commands to coordinate the activity.

❑ Lift and move materials in a manner that avoids personal injury or material damage.

❑ Do not lift materials requiring above average strength or agility.

❑ Never carry an object you cannot see over or around.

❑ Do not manually lift loads that are bulky, unstable, or too large to see over.

❑ Place yourself in a stable position. Be sure your footing is firm. Spread the legs slightly apart to maintain center of gravity.

❑ Point feet, knees, and torso in the same direction when lifting, pulling, or pushing. Turn the whole body when changing direction.

❑ Stand close to the load. Keep the material within 10 inches of your body to reduce strain on your arms and back. The suggested position is elbow height.

❑ Get a secure handhold before lifting. Be sure to grip the load firmly. Place your hands where they will not slip.

❑ Use both hands to grip, lessening the strain on the wrists and forearms. Use gloves if needed.

❑ Lift the material only at those locations designated for lifting or moving. These locations are often designated as "Lift Here."

❑ Avoid getting your fingers, hands, or other body parts pinched between the load and objects nearby.

❑ Squat to lift heavy objects from the floor.

❑ To adjust your grip, set the object down.

❑ Lift with your legs—not with your back.

❑ Straighten up slowly and smoothly, avoiding jerking motions.

❑ Keep your back straight.

❑ Avoid twisting by facing the load directly and tucking in abdomen and buttocks. Avoid overextending or twisting your back.

❑ Breathe in before the lift. Inflated lungs help support the spine.

(continued on next page)

- ❏ Hold the load firmly. Keep it close to your body. Exhale.
- ❏ Move the item to its new location. Move smoothly and avoid rapid and sudden motion.
- ❏ Do not lift or stack the material above shoulder level.
- ❏ To put the load down, bend at the knees. Keep your back as straight as possible.

Storage

- ❏ Make sure the location is not in an area where excessive heat or cold can damage the material.
- ❏ Make sure the stacking location is stable.
- ❏ Store materials and chemicals in approved containers and locations only.

MEETS THE FOLLOWING (MSSC) PRODUCTION STANDARDS	
P1-K7G	Material handling procedures are followed to prevent product damage.
P1-O7D	Knowledge of how to safely move materials.

COPYRIGHT © GLENCOE/MCGRAW-HILL

Using Mechanical Lifting & Moving Equipment

When moving materials, you need to follow both the general rules for mechanical lifting and moving as well as the specific rules that apply to the equipment you are operating.

Equipment

❑ Lift only with the equipment designated for lifting and moving the material. Specific equipment (forklift, hand truck, hoist, or hand or electrically powered dolly) is often identified.

❑ Make sure the equipment is in proper operating condition.

- Do the brakes work?
- Do the wheels rotate freely?
- Are the controls and handgrips in good condition and comfortable?

Travel

❑ Plan your route of travel.

❑ Make sure the route of travel is free of obstacles.

❑ Make sure the surface is clean, even, and free of slippery areas.

❑ Do not make sudden stops, starts, or changes in direction.

Lifting

❑ Before handling unfamiliar and hazardous materials or chemicals, read the label for safety instructions. Refer to the Material Safety Data Sheet (MSDS) or consult your supervisor.

❑ Make sure the equipment is stable before lifting any material.

❑ Do not exceed the rated capacity of the lifting equipment.

❑ Lift and move materials in a manner that avoids personal injury or material damage.

❑ Lift only at areas designated. These areas are often marked "Lift Here" or "Place Forklift Here."

❑ Keep material upright and stable.

❑ Lift material only high enough to clear the floor while moving.

❑ When moving items on dollies or hand trucks, push rather than pull whenever possible.

Storage

❑ Make sure the storage location is not in an area where excessive heat or cold can damage the material.

❑ Store materials and chemicals in approved containers and locations only.

❑ Make sure the stacking location is stable.

❑ Raise material to a stacking position or location only at the stacking area.

❑ Do not stack individual items higher than the designated stacking height.

❑ Make sure the material is stacked evenly and that the stack is stable.

MEETS THE FOLLOWING (MSSC) PRODUCTION STANDARDS

P1-K7G	Material handling procedures are followed to prevent product damage.
P1-O7D	Knowledge of how to safely move materials.

Shipping Products

Shipping products from the manufacturer to the customer involves many key steps. The basic steps are packaging the product, labeling the package, and sending the package. Each one of these steps can be divided into sub-steps. In all cases, regardless of the complexity of the process, follow these general guidelines.

- ❑ Double-check the shipment against the work order.
- ❑ Complete all necessary documentation.
- ❑ Make sure that the shipment complies with all applicable laws.
- ❑ Ship the product on time.

Using the following checklist will help ensure that products are ready for shipment. It will also help guarantee that products arrive undamaged and on time.

- ❑ If there will be a delay in shipment, has this been communicated to the customer and the other proper parties in a timely manner?
- ❑ Has the product been properly stored in preparation for shipping? Proper storage is essential to maintain the quality of the product.
- ❑ Have the product, the quantity, the destination, and the packaging instructions been checked against the work order?

- ❑ Have packaging documentation and customer shipping instructions been completed and placed with the product to accompany it to the next destination?
- ❑ Do the packaging materials meet packaging and shipping specifications, including proper labeling?
- ❑ Has a decision been made on how the product will be shipped?
- ❑ Have all laws and regulations with regard to labeling, packaging, and transport been correctly followed?
- ❑ Have any possible delivery issues been resolved?
- ❑ Is the package sealed properly?
- ❑ Is the package labeled properly?
- ❑ Is the destination address correct and free of errors?
- ❑ Has the package been weighed?
- ❑ Has the correct postage been applied if necessary?

MEETS THE FOLLOWING (MSSC) PRODUCTION STANDARDS

P1-K7	Prepare final product for shipping or distribution.
P1-K7A	Packaging materials meet packaging and shipping specifications, including proper labeling.
P1-K7B	Completed documentation of packaging and customer shipping instructions accompany the product to the next destination.
P1-K7C	Product availability is communicated to the proper parties in a timely manner.
P1-K7D	The product and all relevant information such as quantity, destination, and packaging instructions, are checked against the work order.
P1-K7E	Product is correctly stored or staged for shipping.
P1-K7F	All laws and regulations with regard to labeling, packaging, and transport are followed.
P1-K7G	Material handling procedures are followed to prevent product damage.
P1-O7B	Knowledge of state and federal regulatory requirements (e.g., OSHA).
P5-A16E	Calculate weights of materials and delivery issues.

COPYRIGHT © GLENCOE/McGRAW-HILL

Completing Shipping Documents

Packaging and shipping are the final steps in the production process prior to the customer receiving the product. Because shipping errors lead to customer dissatisfaction, it is important to follow the instructions on shipping orders and picking tickets carefully. In addition to confirming that the correct product is sent to the customer, workers involved in packaging and shipping must pay particular attention to mathematical calculations and units. Shipping errors can add substantial shipping and inventory cost as products are returned and orders are reshipped.

The shipping order specifies:

- The products to be shipped.
- Customer information.
- Shipment method.

As products are packed, they are checked off on a picking ticket, the list of items to be included in the shipment. This list may be a paper document or a list on a computer screen.

Calculating Amount of Product to Ship

Most shipping orders and picking tickets include information about the product identification and the quantity to be shipped. If each item is packed individually, you need to count the number of items placed in the package and enter that number on the picking ticket. However, some products are shipped in packs or cases that contain more than one unit. In this case, you must confirm whether the shipping order calls for a specific number of items or a number of cases. It is important that the customer receive the correct number of products.

Every company has it own forms, so your picking ticket may have information arranged differently than the example shown. However, all of the information should be included. It is important to note the column labeled U/M. This is the unit of measure, and it describes what is being designated by the quantity column. If the unit of measure differs from the units that you are shipping, you need to calculate the number of items to pack. In the final column, you enter the number of units packed in the shipping container, using the same units as designated in the U/M column.

What do you do if you are filling the order and find that item R8872 is packaged in cases of 10? You must determine how many cases to ship—in this example $30 \div 10 = 3$ cases. Pack 3 cases in the shipping container and enter 30 in the units packed column. If you are filling out a paper ticket, you can write "3 cases of 10" instead of the number 30, but the actual number of units must be clearly indicated. For item H4088, you will ship 5 cases and enter the number 5 in the units packed column.

Picking Ticket				
Part number	Location	Quantity	U/M	Units packed
H4088	Bin 64	5	case of 6	
E345	Bin 48A	3	pack of one dozen	
R8872	Bin 48B	30	each	
L998	Bin 8	100	each	

(continued on next page)

APPLYING YOUR KNOWLEDGE

Use the following picking ticket columns to answer the questions.

Picking Ticket				
Part	Location	Quantity	U/M	Units packed
Small gear	Bin 64	30	case of 20	
Large gear	Bin 48A	4	case of 10	
Spring	Bin 48B	2	boxes of 144	
Ratchet assembly	Bin 8	6	each	
Handle	Bin 3	25	each	

1. How many total small gears are to be shipped? What number is entered in the final column?

2. Large gears are stored in boxes containing 2 cases per box. How many boxes will you package?

3. How many springs were ordered?

4. What is the unit by which ratchet assemblies are ordered?

5. You go to Bin 3 and discover that handles are now packaged in boxes of 5. How many boxes do you ship? What number do you enter on the picking ticket for handles?

PRODUCTION CHALLENGE

As the supervisor of the shipping department, you find that a new hire in your department is consistently failing to complete certain documents correctly. This has led to delayed shipments, misdirected shipments, and shipments via the wrong carrier. The new hire has completed your department training program. What would you do to ensure that the shipping documents are completed correctly?

MEETS THE FOLLOWING MSSC PRODUCTION STANDARD

P1-A16E	Calculate the correct amount of products shipped.

COPYRIGHT © GLENCOE/MCGRAW-HILL

Implementing ISO 9000

ISO 9000 is a series of universal standards that defines a quality assurance system. It was developed by the International Organization for Standardization (ISO) and adopted by 90 countries. Like Total Quality Management (TQM) and other quality tools, the ISO 9000 Standards help businesses identify problems and record quality issues. The ISO 9000 Standards are not quality control standards for products. Instead, they are management systems standards. They help businesses ensure that their operations are efficient, effective, and standardized.

The ISO 9000 Standards can be used in very small and very large businesses. The standards do not dictate how businesses must reach those requirements. The method of implementation will differ and will depend on the size and structure of the organization.

If your employer decides to incorporate these standards into their quality assurance system, you might be involved in a team that implements those standards. The following are general guidelines for implementing ISO 9000 Standards.

1. Identify goals as they relate to your quality management system. These might relate to increased efficiency, increased profitability, or increased customer satisfaction.

2. Write down these goals.

3. Obtain the ISO 9000 Standards and related materials. Copies are available from American National Standards Institute, 11 East 42 St., New York, NY 10036.

4. Identify those ISO 9000 standards that relate to the goals.

5. Perform a self-evaluation. Identify the gaps between your company's quality management system and the ISO 9000 Standards.

6. Once you have identified the gaps, draw up a plan to implement the ISO 9000 Standards in an effective, efficient, and responsible fashion.

7. Identify available resources. The effort required to comply with ISO 9000 Standards will depend on the effectiveness of your present quality system. There are training resources to help companies achieve ISO 9000 registration. Check with your local community college, university, or small business development center. They may be able to provide information about classes, recommend consultants, and arrange for preregistration assessments. Small business development centers may offer group classes in ISO 9000.

8. Consider what personnel will be needed to implement the plan. You may need to hire an outside consultant. You might also want to look within your own company. Employees who are familiar with ISO 9000 might be able to assist you.

9. Make co-workers your partners in implementing ISO 9000 Standards. To do this, you will need to:

   ■ Educate co-workers about the ISO 9000 system.

   ■ Make sure they understand why you are implementing it.

   ■ Make sure they understand how you are implementing it.

   ■ Encourage your employees and co-workers to provide feedback on aspects of the present system that do and do not work.

   ■ Work with team members to achieve measurable improvement in quality.

(continued on next page)

10. After implementing the ISO 9000 Standards, perform periodic self-evaluations to ensure that you are following the standards. These assessments will help you target areas for improvement.

11. Decide whether the company should seek registration for your management system. An ISO 9000 registration confirms that conformances to quality standards have been properly assessed and documented. ISO 9000 registration will be vital to your company's success if:

    ■ Your customers require it of their suppliers.

    ■ Your industry sees a strong need for ISO 9000 registration.

    ■ Your competitors are working toward registration.

    ■ Your company could benefit by establishing a formal quality system to improve quality and reduce errors, returns, and customer complaints.

12. You will need to comply with certain requirements. You will be asked to:

    ■ Write a quality manual to describe your quality system.

    ■ Document how work in your organization is performed.

    ■ Design and implement a system to prevent problems from recurring.

    ■ Identify training needs of employees.

    ■ Calibrate measurement and test equipment.

    ■ Train employees on how the quality system operates.

    ■ Plan and conduct internal quality inspections, or audits.

    ■ Comply with other requirements of the ISO 9000 Standards as needed.

13. The registration certificate stating that your company complies with the ISO 9000 Standards can be issued only by an accredited, third-party registration agency. This agency is called the registrar. You should select the registrar at the beginning of the process. You should then find out in detail what they require before they will grant a registration certificate. Typically, the registrar will conduct a preassessment audit to identify areas of noncompliance. This gives you the chance to correct those areas before the registration audit. The charge for the final registration audit is set by the registration agency.

14. The registration agency may conduct audits at six-month intervals to ensure that the company continues to comply with the ISO 9000 Standards. An ISO 9000 registration certificate is valid for three years.

15. Continue to evaluate and improve your business.

MEETS THE FOLLOWING PRODUCTION STANDARDS

P1-040	Knowledge of how to implement quality assurance principles and methods, such as ISO 9000.
P1-A11C	Acquire training on ISO 9000.
P4-K1D	Conformances to quality standards are properly assessed and documented.
P4-01E	Knowledge of quality management systems and how to use them to perform quality checks.
P5-01E	Skill in interpreting quality requirements, industry standards, and documentation requirements.
P8-01A	Knowledge of quality standards and how they apply to products to make effective decisions about quality problems.
P8-02A	Skill in using Total Quality Management (TQM) and other quality tools to identify problems and record quality issues.

COPYRIGHT © GLENCOE/McGRAW-HILL

Calibrating Dial Calipers

Safety First	Tools and Equipment:
❑ Wear safety glasses at all times. ❑ Follow all safety rules when using hand tools and measuring instruments.	■ Dial caliper ■ Gauge blocks: 1.000", 2.000", and 4.000" ■ Surface plate

The calibration of other measuring tools can be checked in much the same way as the calibration of the dial caliper in this example. Checking the calibration of a gauge or any piece of data collection equipment involves comparing a given dimension against a known standard dimension and verifying that the given dimension falls within the specified tolerance range.

Procedures: Refer to the appropriate calibration instruction manual for specifications and special procedures.

1. Ensure that the calibration schedule is implemented according to specifications.

2. Ensure that instrument certification is checked both by reviewing documentation and careful observation during use.

3. Instruments that are out of calibration should be immediately recalibrated or referred to the appropriate parties for recalibration or repairs.

4. Visually inspect the calipers for damage. If there is damage present, have the instrument repaired before proceeding.

5. The calipers should be clean and well lubricated, and all parts should demonstrate the proper freedom of movement for the particular instrument.

6. Ensure that the dial indicator on the caliper reads 0 when the caliper is fully closed. If necessary, adjust the bezel.

7. Using the outside jaws of the caliper, measure the length of the gauge blocks: 1.000", 2.000", and 4.000".

8. Record the results for each gauge block reading in the Reading Before Adjustment column of the calibration record.

9. Reset the jaws to the fully closed position and verify that the dial indicator reads 0.

10. Place the 2.000" gauge block on the surface plate and measure the height of the block using the depth gauge of the caliper.

11. Record the results in the Reading Before Adjustment column of the calibration record.

12. Calipers are considered out-of-calibration if any measurement is outside of a ± .001" tolerance from the accepted block size measurement.

13. If the calipers are within the calibration tolerance, then it is not necessary to indicate a measurement in the Reading After Adjustment column. Simply place a dash "—" or "N/A" to indicate that an entry is not applicable.

14. If the calipers are outside of the calibration tolerance, then it will be necessary to bring them back into calibration by making an adjustment.

15. Often, bringing a measuring tool back into calibration is performed by a supervisor in a quality control-oriented position. Check with an immediate supervisor to learn who is responsible for recalibrating measuring tools and then bring the tool to the attention of that person.

(continued on next page)

Calibration Record				
Name:			Mfg/Model:	
ID#			Calibration Standard:	
Calibration Interval:			Acceptance Standard:	
Calibration Points:				
Date Checked	By	Readings Before Adjustment	Readings After Adjustment	Date Due

Key: **Calibration Standard:** the standard used for checking the tool

Calibration Interval: the frequency of required tool calibration

Acceptance Standard: the tolerance that determines whether or not a tool is in calibration

Calibration Points: the location in a tool's range of measurement where calibration will be checked

MEETS THE FOLLOWING (MSSC) PRODUCTION STANDARDS

P1-O1A	Skill in using inspection equipment, including how to calibrate, what type of equipment to use, and what frequency to use.
P4-K2	Check calibration of gauges and other data collection equipment.
P4-K2A	Calibration schedule is implemented according to specifications.
P4-K2B	Instrument certification is checked both by reviewing documentation and through careful observation during use.
P4-K2C	Instruments that are out of calibration are immediately recalibrated or referred to the appropriate parties for recalibration or repairs.
P4-O4A	Skill in verifying calibration of inspection equipment.
P4-O4F	Skill in maintaining and storing inspection tools.
P4-A2E	Check all inspection equipment.
P4-A15B	Read calibration manuals and be able to implement corrective actions.
P4-A16E	Calculate equipment calibration.

COPYRIGHT © GLENCOE/MCGRAW-HILL

Calculating Safe Volumes of Contamination

A contaminant is something that renders something else impure, inferior, or unfit for use. This is a rather broad definition. In practical use, the definition is limited to the adjectives "inferior" and "unfit." After all, adding carbon to iron renders it impure, but it serves to create an even stronger and more versatile metal that we call steel. We do not generally consider steel to be "contaminated" by carbon.

When a contaminant might render a product or the environment inferior or unfit for use, its presence must be controlled. Different contaminants will make different products inferior or unfit for use when the level of a contaminant exceeds a given tolerance.

Role of the EPA

For example, pesticides are widely used in producing food and beverages. Residues, or very small amounts, of the pesticides may remain in or on such foods as fruits, vegetables, and grains. The Environmental Protection Agency (EPA) regulates how much of each pesticide may remain in or on these foods and sets a maximum residue limit, or tolerance, for each. Tolerances for a given contaminant are determined in a manner similar to the one that follows.

- A study is conducted on how much of the pesticide is regularly used and how often.
- A study is conducted on how much of the pesticide remains on food when it is marketed or prepared.
- A study is conducted on the toxicity of the pesticide.
- A study is conducted on the toxicity of the by-products of the pesticide.

This process is repeated many times over. All of the information from all of the different studies is combined and analyzed. The result is a tolerance that represents the maximum allowable concentration of the pesticide in or on a given product by the time it reaches the consumer.

Role of the FDA

This data is passed on to the Food and Drug Administration (FDA), which creates regulations that manufacturers must follow. The FDA may also test products to ensure that manufacturers are in compliance with the regulations. They follow a process similar to the following.

- A sample of a given product that contains ingredients that may have been exposed to a pesticide is gathered.
- The sample is placed in a special blender along with specialized organic solvents and is then puréed.
- The sample is then divided into its constituent parts using special equipment and more organic solvents.
- Water is chemically removed from the sample.
- A portion of the sample is then placed in a specialized machine such as a gas chromatograph that will analyze the sample and reliably indicate the concentration of a particular pesticide or other contaminant in the sample.

Specific companies may also have their own tolerances and testing procedures for use in quality control. Familiarize yourself with your company's testing procedures and follow them to help ensure that high quality products are shipped to the customer.

MEETS THE FOLLOWING (MSSC) PRODUCTION STANDARD

| P3-A16B | Calculate the safe volumes of contamination. |

Developing Performance Indicators

Performance indicators (PIs) are used in a quality system to monitor how a system is functioning and as an indicator of necessary changes. A PI is a measurement of the performance of a process or a system, so it must use data that can be quantified. Essentially, a PI is a numerical measure of how well you are achieving a goal. In order to write a PI that provides usable data, you must know what you want to measure and why.

The writer must:
- Know the goal.
- Know how the measured value corresponds to that goal.
- Be able to understand and clearly state the purpose and the steps to follow in using the PI.

The operator must:
- Be able to understand the purpose of the PI.
- Know how to use the PI.

The checklist below will help you write a performance indicator that provides useful information and can be understood by the workers who use it.

Planning

1. Determine the needs of the internal and external customer. What information will indicate how well the process is working?
2. Identify which company or department goals will be addressed by the PI. What goals are you trying to accomplish?
3. Determine which aspects of the process can be measured. Is there a quantitative value, such as assembly time, number of products, number of defects, or number of accidents and near misses, which addresses the goals?
4. Select the appropriate PIs and how they will be measured.
5. Determine the target values and tools, such as charts or computer programs, that will be used to track the data.
6. Determine who will collect the data and who will process it. Will all of the information be obtained and evaluated by frontline workers?

7. Determine whether the workers responsible for collecting and using the data have received appropriate training. Are they aware of the purpose and procedures for data collection and handling?

Writing the Performance Indicator

8. State the objectives—clear, measurable, achievable results—of the PI and how it fits into overall goals and objectives of the company or the department. State clearly what you are trying to achieve.
9. Write a statement of the needs to be addressed and who will use the data.
10. Write clear directions for obtaining the measurements, or refer to and attach appropriate standard operating procedures for measurement. Use simple declarative sentences.
11. Indicate what training is necessary in order to perform the data collection or evaluation.
12. Indicate how the data will be used to improve the process or determine whether changes need to be made.

Follow-Up

13. Present the PI to workers involved with the process. Get feedback about how well they understand the procedures.
14. Address any questions raised by workers.
15. Determine whether more training is needed.

MEETS THE FOLLOWING PRODUCTION STANDARDS

| P1-040 | Knowledge of how to implement quality assurance principles and methods, such as ISO 9000. |
| P8-02F | Skill in developing performance indicators that can be readily understood by operators. |

COPYRIGHT © GLENCOE/MCGRAW-HILL

Documenting Quality Processes

Documentation is an integral part of the quality process. Documenting issues and changes helps ensure the future quality of the product. Different manufacturers have different methods for keeping track of quality control data. However, many quality control documentation systems are based on similar principles and methodologies. The method discussed here uses Microsoft Excel™ and Word™ to track, maintain, and report the data. You will need to become familiar with the system your company uses.

Organize Data

Once the data collection, maintenance, and reporting procedures have been established, organize the data in a spreadsheet or database. Consider the following example. A manufacturer produces a variety of microprocessor chips for the computer manufacturing industry. To reduce overall costs, the company has implemented a quality control documentation system. They will use the system to identify and track quality issues before they become problems. You are in charge of organizing the collection, maintenance, and reporting of this data. The following types of data can be collected on any given product. They are shown in **Fig. 1**.

■ **Fig. 1**

Batch Number	00001	
Chip Type	MP	
Physical Dimensions and Tolerances (mm)	26 mm × 22 mm	±0.10 mm

- Batch number. Chips are produced in batches. Each chip is part of a particular batch. For the purpose of this example, the batch number will start at 1.
- Chip type. Only one type of chip is produced in each batch. There are five types of chips (microprocessor MP, logic circuit LC, math co-processor MC, signal amplifier SA, and analog-to-digital conversion processor AD).

- Chip number. It would be impractical to test every single chip, but given the stringent requirements of the product a sufficient number must be sampled to accurately represent the entire batch. To accomplish this, the first chip and every tenth chip thereafter is tested. Batches are 100 chips. See **Fig. 2**.
- Physical dimensions and tolerance. Each chip has physical dimension design specification with a tolerance of ± 0.10 mm.

Document the Process

Using this information, construct a spreadsheet in Excel that documents your inspection of the chips. Once the spreadsheet is constructed, establish a process for documenting all quality-related data. Then set up a system and establish standards for maintaining the records on quality processes. The following is an example of how an Excel spreadsheet could be set up to document a process.

1. Determine which data must be collected, maintained, and reported.
2. Organize the data types that you intend to collect. Information that would be the same for every entry can be entered once in a general information section at the top of the sheet.
3. Set up an organized table into which variable data can be input.
4. Key all the necessary data into the respective fields.

(continued on next page)

5. Use the Excel charting wizard to create a graph that summarizes the important data.

6. Evaluate the data to detect trends that could signal a quality issue.

7. Evaluate your arrangement of the data on the spreadsheet. Is the data easy to read and understand?

8. Correct the arrangement of the data if necessary.

In the example, all of the data that could be recorded about the given product sample was keyed into an Excel spreadsheet, as shown in **Fig. 2**. Consider the following regarding the use of this data.

- You may not need all of the data available. Some data may be superfluous to any required analysis.

- After recording the data, recheck it for accuracy. This maintenance of the data helps ensure that the conclusions drawn from a later analysis are valid.

- When it comes to reporting data, the overall quantity of data expressed declines. If a report were made of the preceding example, it would likely not include each of the individual data points from the samples collected. Rather, it would give an average along with other statistical information that would effectively sum up the data.

- The graphs would be included to provide an at-a-glance look at the quality.

- Each manufacturing situation is different and thus may require a different format, but the basic principles are still the same.

■ **Fig. 2**

Chip Number	Physical Dimensions: Length (mm)	Physical Dimensions: Width (mm)	Pass/Fail
1	26.02	22.06	Pass
11	26.04	22.00	Pass
22	26.01	22.03	Pass
33	26.03	22.02	Pass
44	26.04	22.04	Pass
55	26.06	22.03	Pass
66	26.05	22.05	Pass
77	26.07	22.01	Pass
88	26.07	22.01	Pass
99	26.09	22.00	Pass

(continued on next page)

COPYRIGHT © GLENCOE/MCGRAW-HILL

Analyzing the data is the next step. In this data analysis, note that the graph in **Fig. 3** shows that all of the chips inspected are within spec tolerances. However, there is clearly a trend indicating a possible quality problem. The length of the chips produced displays a trend toward the outside of the spec tolerance limit. If the trend continues, the chips produced will fail the quality inspection. Documenting such a quality problem provides the following advantages:

- It makes it easier to raise and address quality issues in a timely manner.

- Minor adjustments can be made immediately, which prevents parts from going out of spec and producing costly scrap.

- By properly documenting, tracking, and reporting adjustments and issues back to the original communicator, it is possible to make future fixes faster if the problem cannot be eliminated altogether.

Communicating the quality requirements, issues, and training needs in a timely manner is an integral part of the continuous improvement process. Make sure that your communication reflects knowledge of the quality requirements. If there is something that you do not understand, increase your knowledge in that area. This will further your ability to communicate.

■ **Fig. 3**

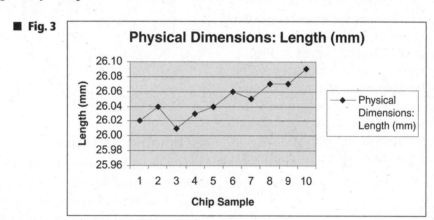

MEETS THE FOLLOWING (MSSC) PRODUCTION STANDARDS

P5-K3	Communicate quality requirements, issues and training.
P5-K3A	Communication reflects knowledge of quality requirements.
P5-K3B	Quality issues are raised in a timely way.
P5-K3C	Quality issues are addressed in a timely way.
P5-K3G	Quality issues are recorded, and tracked and reported back to original communicator.
P5-K4F	Issues are evaluated, tracked and reported back to original communicator.
P5-A14C	Document into quality system the defects in parts produced.
P8-K1C	Quality problems are documented according to established processes.
P8-K2A	Minor quality issues or adjustments are made immediately.
P8-K2B	Quality issues or adjustments are documented properly.
P8-K4A	Records on quality process are maintained to appropriate standards.
P8-K4B	Outcomes of quality processes are charted according to appropriate methods and standards.
P8-K4C	Data on quality process performance is accurate.
P8-K4E	Quality process performance data is reported to appropriate parties in a timely manner.
P8-K5A	Performance and training issues related to quality are identified in a timely manner.

Using Statistical Process Control Charts

Every process includes a certain amount of variation. Depending on the amount and type of variation, the process may be either in control or out of control. When it is in control, the variation is attributed to random causes and when it is out of control, the variation is attributed to specific, or assignable, causes. Statistical process control (SPC) charts are used to analyze the variation within a process. SPC gives frontline workers a tool to determine when a statistically significant change has taken place in the process or when a seemingly significant change is just due to chance causes.

Random Variation

The terms "in control" and "out of control" refer only to whether the process is operating within the range of random variation. They do not correspond to products meeting or failing to meet specifications. As an illustration of this concept, consider two archers, one an expert and the other a novice, at a target range. You would expect the expert's target to look like the target in **Fig. 1**, with all the holes in or near the bull's-eye. The novice tends to have wide variation, as in the target in **Fig. 2**. Even though the scores are very different, both processes are in control, because the variations due to random differences in how the arrow is fired are consistent with past results for the two archers.

■ **Fig. 2**   Novice in control

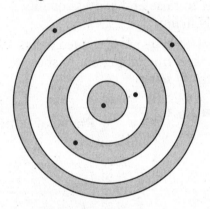

Assignable Variation

Now consider the target in **Fig. 3**, which is the expert's second target of the day. Four holes are in the bull's-eye, but the final shot is well off the mark. Although the target scores much better than that of the novice, it indicates that something was out of control. Perhaps the archer was distracted or had a muscle cramp. In either case, the unusual result can be attributed to an assignable cause and is statistically out of control.

■ **Fig. 1**   Expert in control

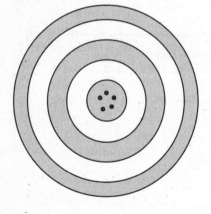

■ **Fig. 3**   Expert out of control

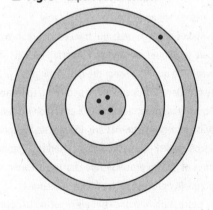

(continued on next page)

COPYRIGHT © GLENCOE/MCGRAW-HILL

Analyzing SPC Charts

SPC charts plot the value for one process parameter and analyze the variation. Key values used to determine whether the process is in control are:

- The center line (CL).
- The upper control limit (UCL).
- The lower control limit (LCL).

Statistical equations, using historical data for the process, are used to determine these values. You will probably not be responsible for determining these values. In most cases, a statistician, engineer, or even a computer program provides them. Frontline workers then use the data to chart the process. If you are required to know how to perform statistical analysis in more depth than discussed in this worksheet, your employer will schedule training, either on site or through classroom instruction.

The parameters that you plot on an SPC chart depend on the process. They can include the amount of time spent on a step, the number of defects in a lot of products, dimension measurements of a part, or anything else about a process that can be measured. Decisions about what to measure, sample size, and sample frequency are based on the process itself.

In Control or Out of Control

Frequent adjustments and changes to equipment do not necessarily improve a process. In fact, adjustments to a process that is in control can be a source of assignable variation, causing the process to be out of control. To determine whether a process is out of control, you analyze the plotted results on the SPC chart. If the process is in control, you continue to monitor it. If it is out of control, you must identify the assignable causes, correct them, and continue to monitor the process to determine whether the change has brought it back into control.

There are several different indications that the process is out of control, based on data falling outside the control limits or process trends that indicate a steady or cyclical change in conditions. The most common tests for assignable variation are:

- 1 data point falling outside the control limits.
- 6 or more points in a row steadily increasing or decreasing.
- 8 or more points in a row on one side of the center line.
- 14 or more points alternating up and down.

For example, examine the charts in **Fig. 4** and **Fig. 5**, which plot process time. The process in **Fig. 4** is in control, because data points are within the control limits, and data points are randomly distributed, showing no trends. The process in **Fig. 5** is out of control, because one of the data points falls above the UCL.

■ **Fig. 4**

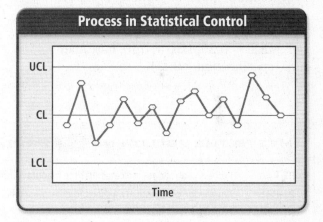

■ **Fig. 5**

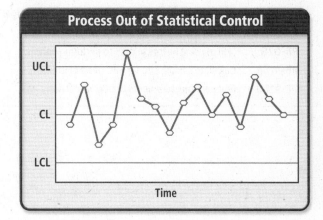

(continued on next page)

APPLYING YOUR KNOWLEDGE

For each of the following processes, create a control chart for the data provided. Determine whether the process is in control and explain why.

1. Run time measurements, in hours: 2, 2.5 , 1.75, 2.25, 2, 1.75, 1.75, 2.25 (CL = 2 h, UCL = 2.75 h, LCL = 1.25 h).

2. Part length in mm: 33.9, 33.0, 33.2, 33.5, 33.7, 33.6, 33.5, 33.8, 33.4 (CL = 33.4 mm, UCL = 33.8 mm, LCL = 33.0 mm, customer specification = 35 ± 0.5 mm).

3. Contaminants in product, in percent: 0.15, 0.19, 0.11, 0.13, 0.14, 0.16, 0.17, 0.18, 0.19, 0.13, 0.19 (CL = 0.15, UCL = 0.20, LCL = 0.10, customer specification <0.17).

PRODUCTION CHALLENGE

Your company produces contact cement. During the production process, the solvent toluene is mixed with an adhesive base to make the contact cement spreadable. How would SPC be used to ensure that this mixing process is in control?

MEETS THE FOLLOWING (MSSC) PRODUCTION STANDARDS

P1-K4A	Process control data indicates that the manufacturing process is in compliance with standards.
P1-O4L	Knowledge of statistical methods to determine when process is out of control.
P1-A14D	Complete SPC chart.
P4-A11A	Take a course on quality tools used by the company (e.g., SPC or statistics).
P4-A11B	Attend classes on SPC.
P4-A16A	Use math to produce charts on department quality levels.
P8-O1F	Skills in evaluating defect patterns.
P8-O2G	Skill in using historical data to perform analysis.
P8-A16A	Calculate sample size for SPC trend of process.
P8-A16D	Compare measurements from control chart data.

COPYRIGHT © GLENCOE/McGRAW-HILL

Selecting a Sampling Plan

Product quality depends in part on product inspection. However, time constraints and costs prohibit the checking of each item in a lot of manufactured goods. Inspecting a small part of the lot is quicker and less costly than inspecting the whole lot. A **lot** is a collection of product units of the same size, type, and style that has been manufactured or processed under essentially the same conditions. The part of the lot that is inspected is called a sample.

Lot Acceptance Sampling Plan

Samples are inspected through sampling plans. The lot acceptance sampling plan (LASP) is one of the most frequently used. The decision is based on the number of defective products in the sample. The decision can be to accept the lot, reject the lot, or even, for multiple or sequential sampling schemes, to take another sample. In that case, the decision process is repeated.

An LASP is a random sampling plan. This is a plan that selects a sample from a lot in which each unit in the lot has an equal chance of being chosen. Random sampling is used to inspect raw materials, items in production, and finished products. The following are the main types of LASP.

Single Sampling Plan. One sample is selected at random from a lot. The quality of the lot is then determined from the resulting information. Such a plan is usually denoted as a (n,c) plan. For a sample size n, the lot is rejected if there are more than c defectives. This plan is the most common and the easiest to use.

Double Sampling Plan. After the first sample is tested, there are three decision possibilities:

- Accept the lot.
- Reject the lot.
- No decision.

If the outcome is No Decision and a second sample is taken, the results of both samples are combined. The final decision is based on that information.

Multiple Sampling Plan. This is an extension of the double sampling plan. More than two samples are needed to reach a conclusion. The advantage of a multiple sampling plan is smaller sample sizes.

Sequential Sampling Plan. This is an extension of the multiple sampling plan. Items are selected from a lot one at a time. After each item is inspected, a decision is made to accept or reject the lot or select another unit.

Skip Lot Sampling Plan. In a skip lot sampling plan, only a fraction of the submitted lots is inspected.

All sampling plans measure the acceptable quality level (AQL). The AQL is expressed as a percentage of products defective, or defects per 100 units. Lots having a quality level equal to a specified AQL will be accepted approximately 95% of the time when the sampling plan prescribed for that AQL is used. To select the appropriate sampling plan, consider the following:

❑ What amount of sampling is necessary? Do consumers of your product require a higher level of quality than your process capabilities indicate?

❑ What is the AQL for the product? The AQL dictates, among other things, the smallest sample size that can be used.

❑ How valuable are the decisions made on the basis of the samples? For a product requiring the highest level of quality, the risk of defect must be very low. To help assure this, the sample size would need to be larger.

(continued on next page)

❏ How serious is the outcome if a defective product is released to the consumer?

❏ Is a large sample size undesirable? If so you will need to reduce the sample size, with the possibility of less accuracy in detecting defects.

❏ What is the cost of sampling and inspecting a lot under various AQLs? The smaller the AQL, the more time needed for sampling and inspecting.

❏ How much time is required to sample and inspect a lot under various AQLs? The smaller the AQL, the greater the cost of sampling and inspection.

❏ Will the costs of the sampling plan be offset by gains in consumer product satisfaction?

❏ What is the record of the quality level of previously submitted lots of the same product? The sample size can be smaller for lots submitted from a production line or supplier with a consistent record of product quality levels significantly better than the specified AQL.

❏ Do the lots consist of product produced under essentially the same conditions? These lots may require smaller sample sizes than those consisting of product produced by different shifts using different raw stock.

MEETS THE FOLLOWING (MSSC) PRODUCTION STANDARDS

P1-K5B	Established sampling plan and inspection policies and procedures are followed.
P4-K4A	Sampling and inspection occur according to schedule and procedures.

COPYRIGHT © GLENCOE/MCGRAW-HILL

Comparing the Product to Specifications

Every product must meet a standard of quality characteristics before it can be released for shipment. A quality assurance inspection measures some quantity of the product, material, or work-in-process sample and compares it to a specification. The specification, which is intended to assure that the product meets the needs of the customer, can be written for any characteristic that can be measured. Often, the specified measurement is a dimension, such as length, measured in units such as centimeter, meter, or foot. Other examples of measurable quantities include weight, volume, number of stitches per inch, or light absorbance by a chemical solution.

Tolerance

A specification always includes a range within which the measurement must fall. This is because there is always some variation both in the product being measured and in the results of the measurement by an instrument. The tolerance is the amount of variation allowed. Tolerance can be expressed as a range, for example:

3.30 mm to 3.60 mm

Any value that falls into that range, including exactly 3.30 mm and exactly 3.60 mm, is considered to be in spec. Tolerance can also be expressed as a range around a central value, for example:

3.45 ± 0.15 mm

Note that this is the same range as 3.30 mm to 3.60 mm:

$3.45 - 0.15 = 3.30$ and $3.45 + 0.15 = 3.60$

The product is in spec if the value falls within the range. For example, if the weight of a container is specified as 500 ± 5 kg and the measured value is 503 kg, then the specification has been met. On the other hand, if the weight of the container is 494 kg, it is out of spec because the value is less than the minimum of 495 kg.

Using Fractions & Decimals

Although the specification sometimes involves only integers—for example, a package of nails could be specified to contain 2,000 ± 2 nails—many specifications will use fractional quantities. These can be expressed as decimal values, as in the example above, or as fractions, such as $\frac{1}{2}$. When you are making measurements, you must be able to work with decimals and fractions to determine whether your measured value falls within the allowed tolerance.

To compare the size of decimal numbers, compare the whole number portions first. The larger decimal number is the one with the larger whole number portion. If the whole numbers are equal, compare the decimal portions. Compare the pairs of digits in each decimal place, starting with the tenths, the first numbers after the decimal point. If that number is identical, continue comparing each successive digit. At the first difference, the number with the larger digit is the larger number. Note that the number with the most digits is not necessarily the largest. For example:

5.21 is larger than 5.2089

because the second digit after the decimal point is larger—1 compared to 0.

(continued on next page)

Example: Specification: 24.50 ± 0.25 cm or 24.25 to 24.75 cm

A measurement of 24.37 cm is in spec.

A measurement of 24.8 cm is out of spec (too large).

A measurement of 24.22 cm is out of spec (too small).

Occasionally a specified value is expressed as a fraction, for example 3 ± ¼ inches. To compare fractions, the denominator (the bottom number) must be the same for each fraction. Would a part measuring 3⅜ inches be on spec or off spec? Change the value ¼ to eighths by multiplying the top and bottom parts of the fraction by the same number.

$$\frac{1}{4} \times 2 = \frac{2}{8}$$

Because 3⅜ is larger than 3²⁄₈, the part is out of spec.

APPLYING YOUR KNOWLEDGE

Determine whether the following measurements pass or fail when compared to the specification and explain your answer.

Measurement	Specification	Measured Value	In Spec or Out of Spec
Length	6.42 ± .05 mm	6.7 mm	1.
Weight	88 ± 1 kg	87 kg	2.
Length	37 ± ⅛ inches	37³⁄₁₆ inches	3.
Light transmittance	48.5%-50.3%	48%	4.
Resistance	1000 ± 50 Ω	978 Ω	5.

PRODUCTION CHALLENGE

You are working on the production line in a plant manufacturing aircraft engines. You have been asked to mentor a new hire at your workstation. This new employee has been trained on the necessary quality control procedures. How would you check that the employee is properly implementing these procedures?

MEETS THE FOLLOWING (MSSC) PRODUCTION STANDARDS

P1-A16C	Measure products against specifications for quality assurance.
P5-K4	Communicate production requirements and product specifications.

COPYRIGHT © GLENCOE/McGRAW-HILL

Calculating Scrap & Rework

In manufacturing, scrap is waste. To a certain degree, material waste is inevitable. However, it can be controlled with some forethought and planning. New technology often comes into play in eliminating material waste. As one example, tool improvements have made manufacturing processes more efficient. This has helped decrease the percentage of waste. It is important to know the amount of scrap produced. The decrease in the percentage of scrap is important in continuous improvement (CI) processes. Only by knowing the percentage of scrap, can you tell whether your efforts to reduce scrap are successful.

The following are among the several ways of measuring scrap.

- Measuring the weight of scrap produced.
- Measuring the volume of scrap produced.
- Mathematically deriving the weight or volume of scrap produced.

Measuring the weight of scrap is by far the easiest. It involves simply placing the scrap on a scale as it is produced and recording the resulting measurement. It is also possible to determine the weight of scrap mathematically, but it is far more practical to measure it directly in most cases. When production is finished, you would add all of the individual scrap measurements to produce the net scrap measurement. This amount can be compared with the amount of bulk material that was initially used. A percentage of scrap can then be figured.

Calculating the Percentage of Scrap

1. Determine the weight of the material scrap.
2. Determine the weight of the original materials.
3. Divide the weight of the scrap by the weight of the original material.
4. Multiply this number by 100 to yield a percentage value. This is the percentage of original material that was lost to scrap.

Measuring the volume of scrap is slightly more complicated. The following are all ways of thinking of scrap volume.

- Physical volume. The actual amount of space that the scrap occupies. The physical volume can be determined through direct measurement or mathematical calculation.
- Unit volume. The quantity of parts relegated to scrap. Unit volume is determined by counting the number of units scrapped.
- Monetary volume. The monetary worth of the scrapped parts and materials. Monetary volume accounts only for the market value of the scrap, though this value is usually tied to a physical unit (i.e., pounds or tons).

Rework is usually expressed as a volume measurement. It might be expressed as the number of out-of-spec units that have to be reworked each day. The calculations for rework are the same as those for scrap.

After calculating the amount of scrap or rework, compile a report that details the situation. Include ideas for process improvement that could reduce or eliminate scrap and rework. Use a word processing program, a spreadsheet program, or a combination of the two to present the data clearly and succinctly.

(continued on next page)

APPLYING YOUR KNOWLEDGE

1. A manufacturer produces gauge blocks for calibrating equipment. Each morning a truck arrives with a delivery of steel bar stock. The shipment consists of 150 steel bars, each with a weight of approximately 49.82 pounds. At the end of each workday, all of the products produced during the day are shipped to customers. On any given day, the total weight of all the products before packaging is 7000 pounds.

   - What is the weight of the scrap produced?

   - What percentage of the total material inflow is lost to scrap on any given day?

   - If steel costs $1.33 per pound, what is the monetary loss of raw materials to scrap?

2. A manufacturer produces gears for use in cars and trucks. Each gear is sold for $5.00 to another local manufacturer. On any given day, assuming that the entire lot of gears is profitable, the manufacturer produces $1750 worth of gears. On average, however, only $1600 worth of finished product leaves the manufacturer. The rest is discarded as scrap. Of the finished product, 6.25% is usually reworked to bring it up to spec.

   - How many gears is the shop capable of producing daily?

   - How many salable gears does the manufacturer typically produce daily?

   - Of the gears that the manufacturer typically produces daily, how many have been reworked?

   - What percentage of possible gears produced daily is lost to scrap?

PRODUCTION CHALLENGE

You are a frontline supervisor in a production facility that manufactures wood shelving for home and commercial use. On being presented with three new home bookcase designs, you notice that one of the designs will require the use of an exotic wood that is available only in nonstandard sizes. These nonstandard sizes prevent full use of all of the material. What would you do to support your argument that the design would be more cost effective if it were modified to specify a wood available in standard sizes?

MEETS THE FOLLOWING PRODUCTION STANDARDS

P5-A16A	Calculate scrap and rework data for reports.
P8-A7A	Demonstrate receptivity to new ideas for CI team to decrease scrap rate.

COPYRIGHT © GLENCOE/MCGRAW-HILL

Reading & Interpreting a Dial Indicator

In general, the greater the degree of precision desired in a finished product, the more often the product must be inspected during production. Frequent inspection is necessary to ensure compliance with standards. The inspection tool used will depend upon what type of measurement needs to be taken and how accurate and precise the measurement needs to be. For instance, if the length, width, and depth of a rectangular block of steel need to be measured within 0.001", you might use a dial caliper to perform the measurement. If a greater degree of accuracy is needed, say 0.00001", then a micrometer might be used. Some manufacturing processes require inspection far more frequently than others.

The dial indicator is one type of inspection tool. It is a measuring instrument that uses a swing-needle and a spring-loaded plunger (movable arm) to display the measured value. The plunger has a specific range (e.g., 1"). The needle on the dial face moves in relation to the movement of the plunger. As the plunger moves, the needle shows the distance or the variation. Readings may be in thousandths of an inch or hundredths of a millimeter.

Tools & Equipment:
- Dial indicator
- Gauge blocks, planer gauge, or any appropriate gauging device

Safety First
- ❑ Wear safety glasses at all times.
- ❑ Follow all safety rules when using hand tools and measuring instruments.

Procedures: Refer to the appropriate instruction manual for specifications and special procedures. Then take a reading, using the workpiece provided by your instructor.

1. Follow all procedures in the appropriate instruction manual.

2. Set the dial indicator to the proper gauging height. The gauging height is the basic dimension of the part to be gauged.

3. Establish the gauging height using gauge blocks, a planer gauge set to the proper height, or any appropriate gauging device.

4. Set the dial indicator gauge at 0. Loosen the bezel lock screw and rotate the bezel (the dial face) so that the 0 on the dial lines up with the indicator arm.

5. Tighten the bezel lock screw.

6. Make sure that the dial indicator is under sufficient spring tension to allow the gauge to rotate in either direction through the desired measuring range.

7. Bring the dial indicator into contact with the workpiece.

8. Read the difference between the gauging height and the height of the workpiece directly from the dial indicator.

9. Record the reading.

MEETS THE FOLLOWING PRODUCTION STANDARDS

P1-01A	Skill in using inspection equipment, including how to calibrate, what type of equipment to use, and what frequency to use.
P1-02A	Skill in reading and interpreting gauges (i.e., analog, digital, and vernier).
P1-04E	Knowledge of how to use and interpret measurement devices.
P4-04E	Skill in using hand-held inspection devices to examine materials.

Conducting Quality Audits

A quality audit is an inspection to determine whether management has defined, documented, and implemented an effective quality system. A quality audit is not a product inspection.

An internal quality audit should be performed in accordance with company and other schedules and procedures. The following are the seven main steps in conducting an audit.

1. **Verify that a quality system policy has been defined and documented.**

   ❑ Prior to the inspection, ask for the written quality policy. Review this document.

   ❑ Check that the policy states the intervals at which audits should be conducted.

   ❑ Check to see whether the quality policy is specific to one product or generic to all products manufactured at the firm.

   ❑ Check to see whether the quality policy is specific to processes or overall systems.

2. **Verify that a quality policy and objectives have been implemented.**

   ❑ Conduct audits in a nonthreatening manner by explaining the reason for the audit.

   ❑ Ask employees whether they are familiar with the quality policy. Do not ask this when the employee is engaged in the actual performance of his or her duties. It could be done when the employee is on break or has finished a task and has yet to begin another task.

   ❑ Check the ways in which management has made the policy available. Is it in the quality manual or another part of the written procedures? Is it posted at points throughout the building? Personnel must know that there is a quality policy and where they can read it.

   ❑ Review employee training records to check employee training in the firm's quality policy and objectives. In particular, this should be done for those employees involved in key operations.

   ❑ When appropriate, the audit should include observation of an operation to ensure that performance meets specs.

3. **Review the company's established organizational structure to confirm that it includes provisions for authorities, responsibilities, and necessary resources.**

   ❑ Review the organizational charts.

   ❑ Check that the company's procedures identify the departments or individuals responsible for performing certain tasks governed by their quality system. They should also include provisions for resources and designating a management representative.

   ❑ Determine whether personnel involved in managing, performing, or assessing work affecting quality have the authority needed to perform designated tasks. Such authority should relate to the reduction, elimination, or prevention of quality nonconformities.

   ❑ Ask the management representative how resources relating to the achievement of quality objectives are obtained and allocated. This will ensure that adequate resources are available.

(continued on next page)

COPYRIGHT © GLENCOE/MCGRAW-HILL

4. **Confirm that a management representative has been appointed and that the appointment has been documented.**

   ❏ Review the company's organizational chart(s) or quality manual to determine whether there is a documented management representative.

   ❏ Determine whether the appointed management representative actually has the responsibility and authority. To determine this, find out:

   ▪ Whether the person has sign-off authority for changes to documents, processes, and product designs.

   ▪ Whether those conducting quality audits report to or provide the representative with their results.

   ▪ How the representative interacts with corrective and preventive actions, relative design control issues, complaints, and product failures.

   ❏ Verify that the management representative is reporting back to management with executive responsibility on the performance of the quality system.

5. **Verify that management reviews, including a review of the suitability and effectiveness of the quality system, are being conducted and documented.**

   ❏ Review the firm's management review schedule to confirm that management reviews are being conducted with sufficient frequency.

   ❏ Check that management review procedures include a requirement that the results of the reviews be documented and dated.

   ❏ Check that management with executive responsibility attends the reviews.

6. **Verify that quality audits, including re-audits of deficient matters, of the quality system are being conducted.**

   ❏ Review the company's quality audit schedules to assure that quality audits are being conducted with sufficient frequency. The time between quality audits should not exceed twelve months.

   ❏ Check that quality audits are conducted by appropriately trained individuals using adequate detailed written procedures to check all elements in the quality system.

   ❏ Critically review the written audit procedures if significant quality system problems have existed both before and after the firm's last self-audit.

   ❏ Check that the audit procedures cover each quality system and are specific enough to enable the person conducting the audit to perform an adequate audit.

   ❏ If it is possible to interview an auditor, ask how the audits are performed, what documents are examined, and how long audits take.

   ❏ Check that audits are conducted by individuals not having direct responsibility for matters being audited. Failure to have an independent auditor could result in an ineffective audit.

   ❏ Check the written audit procedure for instructions for review of audits by upper management. The procedures should require quality audit results to be included in the management reviews.

   ❏ Verify that the procedures contain provisions for the re-audit of deficient areas if necessary.

   ❏ Check that a re-audit report verifies that the recommended corrective action was implemented and effective.

(continued on next page)

7. **Evaluate whether management with executive responsibility ensures that an adequate and effective quality system has been established and maintained.**

   ❑ Check whether your findings indicate that management is appropriately carrying out responsibilities for providing adequate resources and overseeing the quality system to detect problems and address them.

   ❑ Determine whether the management representative and management with executive responsibility are ensuring the adequacy and effectiveness of the quality system.

   ❑ Determine whether that quality system has been fully implemented at the company.

   ❑ Determine whether the effectiveness of the quality policy was evident during the inspection.

   ❑ Properly assess and document conformances to quality standards.

   ❑ Cite any deficiencies showing that major nonconformances indicate that management with executive responsibility is not ensuring the establishment and maintenance of an adequate quality system. This action should be used in those situations where major portions of a quality system have not been established and maintained or whenever there is a total lack of a quality system.

   ❑ Communicate results of audits to employees in a tactful way to bring awareness of areas for improvement.

   ❑ Complete all audit forms in a timely manner.

   ❑ Check that all audit data is relevant and correct.

   ❑ Forward audit forms to the correct parties.

   ❑ Recommend more frequent audits if the firm has a serious quality problem.

   ❑ Conduct your final discussion with management.

MEETS THE FOLLOWING (MSSC) PRODUCTION STANDARDS

P4-K1	Perform periodic internal quality audit activities.
P4-K1A	All audit forms are completed correctly in a timely manner.
P4-K1B	Forms are forwarded to the correct parties.
P4-K1C	Audit data is relevant and correct.
P4-K1D	Conformances to quality standards are properly assessed and documented.
P4-K1E	When appropriate, audit includes observation of operation to ensure performance meets specifications.
P4-K1F	Audit is performed in accordance with company and other required schedules and procedures.
P4-A4D	Determine when and where to inspect or audit process for quality of product or process to meet customer requirements.
P4-A6A	Conduct audits in a non-threatening manner by explaining the reason for the audit.
P4-A6D	Communicate results of audits to employees in a tactful way to bring awareness of areas for improvement.

COPYRIGHT © GLENCOE/MCGRAW-HILL

Completing a Quality Audit Form

A major part of a quality plan that meets the standards of ISO 9000 is the quality audit. The purpose of the quality audit is:

- To determine how well the actual process follows the written standards of the quality plan.
- To identify and report performance and training issues affecting quality.
- To identify potential quality improvements.

A team of auditors can include frontline workers, supervisors and/or managers. It cannot, however, include anyone who does the process or directly supervises the process. You may be involved in quality audits in your own department or other parts of the company.

Before you begin the audit, you will receive an audit form with questions designed to indicate how well the process meets the standards. In some cases, an audit form will already exist, and you will receive the list of specific questions before starting the audit. If no audit form exists for the process, your team may start by writing the form. In this case, you will receive a copy of the quality standards to guide you in designing the audit. The answers to the questions provide data that will be used to evaluate how well the ISO or other quality standards are being applied. They will also be used to determine the effectiveness of the quality procedures, policies, plans, and instructions

Following is part of a checklist for auditing the material receiving process. See the sample audit form on p. 258. A full audit form may be many pages long and is written for the specific process being audited.

❑ To start the audit process, study the audit questions so that you know what information you will be collecting. In order to answer the questions, you need evidence. Evidence is gathered by:

- Interviewing workers.
- Examining process data.
- Observing activities and working conditions.
- Reviewing manuals, documents and records, including health and safety procedures and training records.

As an auditor, you can request any documents or information that you need to complete the form.

❑ As you collect data, answer each audit question and record your observations as notes in the final column. The questions have been designed to provide a complete overview of standards compliance, so every line must have an answer.

- A YES answer means the process complies with the standard.
- A NO answer means that it does not comply.
- A NA answer means that this question is not applicable in the situation.
- Additional observations and explanations for NO answers are written in the OBSERVATIONS column.

(continued on next page)

	MATERIAL RECEIVING AUDIT CHECKLIST	YES	NO	NA	OBSERVATIONS
1.	Is there an established written procedure for the reception of materials and subassemblies?				
2.	Are reception inspections done to all incoming materials and subassemblies arriving at this facility?				
3.	Are purchase orders or any other types of material procurement documents available to the Receiving Inspector to facilitate confirmation of compliance?				
4.	Do the materials and subassemblies receiving inspection confirm and document the following: a. _____ Proper grade of material? b. _____ Proper material markings and identification? c. _____ Proper material dimensions? d. _____ Compliance with dimensional tolerances?				
5.	Are receiving inspections documented for: a. _____ Acceptance and rejection of nonconforming materials and subassemblies? b. _____ Corrective actions taken to deal with non-correctable and correctable nonconformities observed during the reception inspection?				
6.	Are acceptance standards and acceptance tolerances available for reference at the receiving inspection station?				
7.	Does the Fabricator or Manufacturer have a material identification system to assure control of materials of different grade and size (as applicable)?				
8.	Does the Fabricator or Manufacturer segregate controlled materials by project?				
9.	Are materials stored or stocked so as to prevent damage to the raw materials or final fabricated pieces?				
10.	Are the stored or stocked materials clearly marked or identified?				

APPLYING YOUR KNOWLEDGE

1. What would be the best way to obtain information to answer question 1 on the form shown?

2. What would be the best way to obtain information to answer question 4 on the form shown?

3. How would you answer question 8, if the area being audited works on only one project?

PRODUCTION CHALLENGE

You are working in the glass manufacturing industry. You find that you will need to hire consultants to help you conduct an internal quality audit. What would you do to ensure that the consultants have the skills needed to conduct the audit? How would you handle a situation in which an auditor discloses that a personal conviction would affect his or her ability to work on an audit and report certain findings impartially?

MEETS THE FOLLOWING (MSSC) PRODUCTION STANDARDS

P1-040	Knowledge of how to implement quality assurance principles and methods, such as ISO 9000.
P4-A14C	Complete audit form.
P8-K5	Identify and report performance and training issues affecting quality.

COPYRIGHT © GLENCOE/MCGRAW-HILL

Documenting Inspections & Corrective Actions

Safety inspections are often required by company policy or federal regulation to ensure that critical equipment is functioning and in good operating order. These safety inspections are important because they can often catch problems before they become dangerous to workers. Corrective actions are then taken to eliminate the problem and thus maintain a safe workplace. Part of conducting safety inspections is documenting the inspection and any corrective actions that were performed as a result of having performed an inspection. The following is an example of a type of safety inspection that is conducted in the workplace and the documentation for such a process.

Fire Alarm System Inspection

Each workplace must have a fire alarm system, and each system must be periodically inspected to ensure that it is working properly and will function as specified in the event of an emergency. The system is to be maintained according to standard practices. Some parts of the inspection are performed by the end user while others must be performed by a trained professional.

Weekly Inspections

❏ At least one detector, call point, or end-of-line switch on one circuit should be operated to test the ability of the control and indicating equipment to sound the alarm.

❏ If accessible to visual examination, the batteries in the alarm system must be inspected to ensure that they are in good condition.

The following tests must be conducted by trained professionals.

Quarterly Inspections

❏ Batteries must be tested as specified by the supplier.

❏ Alarm sounder must be tested by operation of a detector or call point in each zone.

Annual Inspections

❏ Each detector should be checked in line with the manufacturer's instructions.

Five-Year Inspections

❏ Wiring must be tested in accordance with wiring regulations and all defects logged and rectified.

To ensure that this basic system is functioning, the following routine tests are required:

Monthly Testing

❏ Every detector is to be operated by use of the testing facility (usually the test button on the detector).

Annual Testing

❏ Detectors are to be tested to ensure that they respond to smoke. Open flames, smoke, or aerosols that may damage or affect the sensitivity of the detector should not be used. Use only a suitable test aerosol.

Servicing

❏ Detectors are to be cleaned periodically in accordance with the manufacturer's instructions.

(continued on next page)

FIRE ALARM SYSTEM: Record of Test				
Date	**Call Point/ Detector Location or Number**	**Satisfactory Yes/No**	**Remedial Actions Taken Yes/No**	**Signature/ Name of Company**
January 5, 2005	*CP-203*	*Yes*	*No*	*Jim Jones/ ABC Inspections*
January 12, 2005	*D-06*	*No*	*Yes: Detector battery replaced.*	*Jim Jones/ ABC Inspections*

MEETS THE FOLLOWING (MSSC) PRODUCTION STANDARDS

P3-A14C	Document clear procedures for safety practices.
P3-A14D	Document equipment safety checks in safety log book.
P3-A14F	Document corrective actions regarding safety.
P5-O3C	Knowledge of company reporting forms and documents and procedures specific to safety.
P7-A4F	Check to see if job is safe; if it is not safe, make necessary changes and document why changes were made.

COPYRIGHT © GLENCOE/McGRAW-HILL

Using Quality Circles

Most companies have some type of systematic way of managing quality problems. One common way of addressing quality issues is to form quality circles. A quality circle is a group of people who agree to meet regularly to discuss any quality problems and develop solutions to those problems.

Organizational Structure

Many people are involved in a quality circle. There is a hierarchy that extends from upper management to the workers on the manufacturing floor. At the core of the quality circle are the members. Members are essentially a group of people who work in a similar area and capacity who agree to meet in conjunction with other members of a quality circle to discuss quality problems.

The quality circle includes a steering committee. The steering committee is usually made up of senior executives and top management personnel. This group has responsibility for directing the quality circle, establishing plans and policies, and developing an overall program of operation.

The quality circle also includes the following: a coordinator, a facilitator, and a circle leader. The coordinator organizes the efforts of facilitators and helps direct the overall progress of the quality circles. The facilitator organizes the efforts of several quality circles and may also be called upon to serve as a catalyst for discussion or as a trainer should the need arise. The circle leader oversees the quality circle's activities. The circle leader is often chosen from among the ranks of the circle members.

Quality Circle Tools

There are many tools available to quality circles to help them identify and address quality issues. Among the most commonly used are brainstorming, Pareto diagrams, and cause-and-effect diagrams. The Pareto diagram is a type of bar graph that is used to arrange information in order to prioritize quality problems. The cause-and-effect diagram (also called the Ishikawa diagram or fishbone diagram) is used to graphically present possible causes for a problem. These diagrams are useful in helping to identify trends.

Different organizations may require the use of different tools to accomplish the same task. When choosing tools for analysis, be aware of your company's requirements. Choose tools that fit those requirements that will accomplish the job as quickly and efficiently as possible. Learn to use any diagrams and technical drawings before convening in quality circles. This will help speed up the process.

The following step-by-step process illustrates how quality circles are used with Pareto diagrams, cause-and-effect diagrams, and brainstorming to solve a potential quality issue.

1. Organize a quality circle. Upper management should form a steering committee and nominate a coordinator and a facilitator. Members should form a committee and nominate a leader.

2. Agree on a time to meet and discuss quality issues. Formulate topics of discussion prior to the meeting.

3. At the meeting, group all of the quality issues into categories.

4. Construct a Pareto diagram to help identify areas in which efforts to develop a solution should be concentrated.

Constructing a Pareto Diagram

❏ Determine the category of comparison for the available data in each problem category.

❏ For each category, total the raw data. With the total for each category figured, determine the grand total for all categories.

❏ Organize the categories so that they run from largest to smallest from left to right.

❏ Determine the percentage of problems that each category represents. To do this, sum the totals of all the categories. Divide each individual category by the total and multiply by 100 to get a percentage value.

(continued on next page)

❑ Create a graph. Label the left y-axis with the unit of comparison. Label the right y-axis with percentage ranging from 0 to 100 percent with 100 being equal to the grand total for all the categories. Label the x-axis with the categories.

❑ Draw in bars for each of the categories beginning with the largest. Each bar should represent the total for each category.

❑ Draw a line graph starting at the right-hand corner of the first bar to represent the cumulative percent for each category.

❑ Analyze the chart to determine where corrective efforts should be concentrated. Usually the top 20% of categories will contain 80% of the problems. Such problems can usually be addressed successfully in a systematic manner.

5. Construct a cause-and-effect diagram and brainstorm possible causes.

Constructing a Cause-and-Effect Diagram

❑ Draw a horizontal line. To the right of the line, list the problem for which a cause is to be determined.

❑ The horizontal line is the main or primary "bone" of the fishbone diagram. Extending from the primary "bone" to the left at approximately 45° angles are between two and six secondary bones. Half are on top of the main bone and half are on the bottom.

❑ Label each one of the secondary bones with broad cause categories.

❑ Begin brainstorming specific causes for each of the broad categories and place these on tertiary bones that are connected to the secondary bones.

❑ If necessary, the tertiary bones can be expanded into additional bones to further define the potential cause.

❑ Take care to identify causes rather than symptoms.

6. After identifying problem areas and possible causes, brainstorm potential solutions to the problems. Narrow down the list of solutions to two or three. Build consensus on which solution will ultimately be implemented.

7. Implement the agreed-upon solution. Take the knowledge gained in the process and use it to help prevent future quality problems.

Quality is the responsibility of each individual, from the worker on the floor to the CEO in the office. The following guidelines impact the work of a quality circle. All of these guidelines will help build quality in the workplace.

❑ Individuals should work collaboratively to complete all steps of the production process and suggest improvements.

❑ Issues should be evaluated, tracked, and reported back to the original communicator.

❑ Production workers should be able to adapt their behavior to accommodate quality requirements on different lines and with different products.

❑ Shift-to-shift logs should be completed and problems identified.

❑ Write a brief weekly report to highlight quality problems.

❑ Workers on the floor should identify defect patterns in products and alert management.

❑ Management should respond in a timely manner and work with employees on the floor to eliminate potential quality problems.

(continued on next page)

COPYRIGHT © GLENCOE/MCGRAW-HILL

MEETS THE FOLLOWING (MSSC) PRODUCTION STANDARDS

P1-O6D	Knowledge of how to use diagrams and technical drawings.
P1-A2C	Use quality circles to discuss day-to-day production issues and solutions.
P1-A8C	Work collaboratively with other operators in order to complete all steps of production process.
P2-A14E	Complete shift-to-shift logs.
P4-K5D	Correct analytical tools are selected and used properly.
P4-A7C	Adapt behavior to accommodate quality requirements on different lines and with different products.
P5-K2E	Material and delivery issues are evaluated, tracked and reported back to original communicator.
P5-K4F	Issues are evaluated, tracked and reported back to original communicator.
P7-A9C	Influence others by following safety procedures, making suggestions for improvements and doing a good job.
P8-K4D	Quality process performance data is analyzed to identify trends.
P8-O1K	Skill in identifying inaccuracies in quality data and responding to them.
P8-O4I	Skill in developing and applying preventive actions and mistake proofing.
P8-O4J	Skill in identifying trends that require a systemic solution.
P8-O5B	Skill in analyzing technical data and drawings and gaining group consensus to avoid future non-conformance.
P8-A2D	Analyze the summary and trend information to define the issue or problem.
P8-A3D	Identify defect patterns detected on floor in order to have them corrected by management.
P8-A14A	Write a weekly report to highlight quality problems for the crew.

COPYRIGHT © GLENCOE/McGRAW-HILL

Preventing Electrostatic Discharge

Safety First

❑ Wear safety glasses at all times.

❑ Follow all safety rules when using hand tools and measuring instruments.

Tools and Equipment:

■ ESD wrist strap

■ ESD workstation mat

Electrostatic discharge (ESD) refers to the sudden release of a static electrical charge accumulated on a relatively nonconductive surface. Unlike the current that flows through the electrical conduits of appliances and electronics, static electricity simply accumulates on nonconductive surfaces when the relative humidity is low. The charge is produced when moving materials excite electrons through friction. Over time a considerable amount of charge can accumulate on an object. When the object comes in contact with a conductive material, the whole of the charge is released. This produces the "static shock" that so many of us are familiar with.

The ESD of a static shock can easily damage sensitive electronic parts. The discharge of electricity that we experience when touching a doorknob in the winter is at least 3,000 volts. Any discharge below 3,000 volts is undetectable by human senses. The discharge needed to damage most electronic devices falls somewhere in the neighborhood of 1,000 volts. However, some devices are sensitive enough to be damaged by a 10-volt discharge. The simple act of walking across a carpeted floor can generate between 1,500 and 35,000 volts of ESD. This is an accumulation of electrical charge well in excess of that needed to damage electronics. It is easy to see why it is so important to use ESD-diminishing devices such as the ESD wrist strap

and the ESD workstation mat when working with electronic devices.

Procedures: Refer to the appropriate instruction manual for specifications and special procedures.

❑ Fasten an ESD wrist strap securely to your wrist.

❑ Clip the ESD wrist strap to an ESD workstation mat that has been properly grounded.

Or

❑ Clip the ESD wrist strap to an unpainted grounded surface on the workstation that holds the workpiece.

Or

❑ By means of a special plug, connect the ESD wrist strap to a three-prong grounded wall outlet.

❑ When transporting sensitive electronic devices, use specially designed antistatic packaging.

❑ If possible, keep the relative humidity of the work environment in excess of 40%.

❑ Using an ionic blower will also help reduce the potential for building up an electrostatic charge.

MEETS THE FOLLOWING (MSSC) PRODUCTION STANDARD

P2-A17D	Understanding why an ESD strap must be correctly grounded at the workstation.

COPYRIGHT © GLENCOE/MCGRAW-HILL

Troubleshooting Equipment

Troubleshooting is investigating the cause of a problem and identifying and documenting the appropriate corrective action. Responsibility for troubleshooting varies within an organization. Troubleshooting policies vary from one company to another. Equipment operation manuals usually include information on troubleshooting equipment problems. Those troubleshooting instructions will refer to that specific item of equipment. You might also visit the manufacturer's Web page for troubleshooting tips. Keep in mind that troubleshooting can be necessary on both old and new equipment.

Troubleshooting has three main steps:

- Gathering information.
- Taking corrective action.
- Testing and documenting.

Following is a general procedure for troubleshooting an equipment problem.

Gathering Information

1. Identify the problem.
2. Determine if there is a safety issue.
3. If there is a safety issue, hit the emergency stop button.
4. Call the supervisor.
5. Present the supervisor with the information relating to the problem. At what point did the problem occur? Has the problem occurred before?

Taking Corrective Action

6. Determine whether the problem is a quick fix.
7. If the problem is a quick fix, the supervisor will direct maintenance technicians to correct the problem.
8. If the problem is not a quick fix, the supervisor needs to determine the nature of the problem.
9. If the supervisor determines that the problem relates to setup, the problem will then be corrected.
10. If the problem relates to maintenance, the supervisor will need to generate a work order.
11. If correction of the problem takes priority, the supervisor will need to find out if there is time to correct the problem.
12. If there is time to correct the problem, it should be corrected then.

Testing & Documenting

13. The equipment should then be checked to verify that the problem has been corrected.
14. If further corrective action is needed, that should be taken.
15. The work area should be cleaned up.
16. The work order should be closed out.
17. If the problem does not take priority, the supervisor must deal first with the first-priority problem.
18. If there is not time to fix the problem, the supervisor will need to move the operator to another item of equipment and adjust priorities.
19. The problem and the corrective action should be documented according to company policy.

MEETS THE FOLLOWING (MSSC) PRODUCTION STANDARDS

| P1-O4M | Skills in troubleshooting process to isolate the cause of the problem. |
| P2-A14B | Write reports on troubleshooting results on new equipment. |

Taking Closed-Loop Corrective Actions

Closed-loop corrective actions are performed within a closed-loop system. A closed-loop system is a control system with an active feedback loop. A control system is a system for controlling the operation of another system. Feedback is the process in which part of the output of a system is returned to its input to regulate its further output. A loop is a network in which components are connected in a loop.

Use of closed-loop corrective actions can change the way in which a manufacturer identifies, corrects, and documents a problem. A closed-loop corrective action program requires correct reporting, analysis, implementation, documentation, and verification. This process increases the efficiency of a system. You may need to facilitate a corrective action team to determine what will be required to improve the system and its process.

Reporting

❏ Make sure that you have in place the infrastructure to detect and report existing or potential problems, nonconformities, hazards, and unsafe conditions.

❏ Ensure that the system for failure reporting is consistent among all the parties involved.

❏ Formally report all failures to the appropriate parties in a cooperative and courteous manner.

❏ Write recommendations for corrective action. Complete a corrective action report (CAR).

❏ Communicate all changes related to corrective actions to the appropriate parties, including production operators.

Analysis

❏ Specify a time limit for the completion of failure analysis.

❏ Complete failure analysis within the specified time frame.

❏ Ensure that the system of failure analysis is consistent among all the parties involved.

❏ Analyze all failures to sufficient depth to identify the underlying root cause of failure and the necessary corrective actions.

❏ Conduct failure analysis on every failure, not just on those occurring when repetitive failures occur.

❏ Analyze random failures and test equipment failures.

❏ Consider the history of previous failures.

❏ Consider temperature and other environmental conditions.

❏ Undertake engineering analysis on all engineering failures.

❏ Undertake laboratory analysis on all critical failures.

❏ Do not consign failures to a failure backlog for laboratory analysis.

Implementation

❏ Consult the appropriate parties, including production operators, regarding corrective actions.

❏ Make sure that the corrective action system complies with current regulations.

❏ Ensure that the system of implementing corrective action is consistent among all parties involved.

❏ Implement corrective actions promptly in a consistent fashion and according to company procedure.

(continued on next page)

COPYRIGHT © GLENCOE/MCGRAW-HILL

❑ Implement the corrective action system at all levels throughout the system life cycle.

❑ Ensure that corrective action is not confined to the production phase.

❑ Follow through properly on corrective actions.

❑ Provide feedback and suggestions when corrective actions fail.

Documentation

❑ Integrate document management to provide a comprehensive and accessible electronic or paper trail of changes in the system.

❑ Use centralized data collection to gather information from multiple sources across a product's entire life cycle.

❑ Collaborate with fellow employees on the exchange of information regarding timely corrective action.

❑ Document evidence of the corrective action in a timely manner.

❑ Communicate evidence of the change resulting from the corrective action to the appropriate parties in the correct format.

❑ Provide access to complete records and data for problems, changes, customer complaints, and solutions.

❑ Use the documentation to identify the root of a problem.

❑ Use the documentation to track any trends or changes in the system.

❑ Use the documentation to prompt continual product improvement by identifying trends that require corrective action.

❑ Use process documentation to help manufacturers meet quality requirements.

Verification

❑ Verify implementation of the corrective action through spot checks.

❑ Perform ongoing audits to optimize the outcomes of the corrective action.

❑ Monitor equipment to ensure that corrective action solved the problem.

(continued on next page)

MEETS THE FOLLOWING (MSSC) PRODUCTION STANDARDS

P2-K2D	Equipment is monitored to ensure that the corrective action solved the problem.
P2-K4D	Corrective action is taken to correct unsafe conditions.
P2-O3D	Skill in documenting repairs, replacement parts, problems and corrective actions to maintain log to determine patterns of operation.
P3-K1B	Corrective action is taken to correct potential hazards.
P3-K3A	Conditions that present a threat to health, safety and the environment are identified, reported, and documented promptly.
P3-K3B	Corrective actions are identified.
P3-K3C	Appropriate parties are consulted about corrective actions.
P3-K3D	Corrective actions are taken promptly according to company procedures.
P3-K3E	Ongoing safety concerns are tracked and reported until corrective action is taken.
P3-O4B	Knowledge of required corrective action procedures.
P6-A14C	Prepare written recommendations on internal work flow improvements.
P8-K1A	Quality problems are reviewed with production operators.
P8-K1D	Defect trends are summarized and reported to appropriate parties.
P8-K6	Implement closed-loop corrective action.
P8-K6A	Evidence of corrected action is documented in a timely manner.
P8-K6B	Change resulting from the corrective action is communicated to appropriate parties in the correct format.
P8-K6C	Implementation of the corrective action is verified through spot checks.
P8-K6D	Reports are stored properly for the specified timeframes.
P8-K6E	Ongoing audits are performed to optimize the outcomes of the corrective steps.
P8-O4E	Skill in determining corrective action.
P8-O5A	Skill in facilitating a corrective action team to determine what will be required to improve the system and its process.
P8-A6B	Contact respective parties in a cooperative way to come up with timely corrective actions.
P8-A6C	Contact supervisor in a courteous way in order to report corrective actions.
P8-A14B	Write recommendations for corrective actions.
P8-A14D	Document corrective actions taken.

 COPYRIGHT © GLENCOE/McGRAW-HILL

Completing a Nonconformance Report

Customer satisfaction with a manufactured product depends on the manufacturer's delivery of a defect-free product. If quality is to be guaranteed, manufactured products must be carefully checked against the correct specifications. This will help ensure that they conform to the customer's specifications. Products that do not meet specifications are identified as nonconforming. Such products are not in compliance with the customer's specifications.

Documentation of nonconforming products specifies the following:

- Means for documenting nonconforming products.
- Methods and responsibilities for identifying nonconforming products.
- Individual authorized to evaluate nonconforming products.
- Methods and responsibilities for segregating nonconforming products.
- Method for notifying customers of nonconforming products, if required, and obtaining their approval for use or repair.
- Method of notifying affected groups about nonconforming products.
- Individual responsible for verifying that products conform to requirements after repair or rework.
- Individual authorized to dispose of nonconforming products.
- Individual responsible for determining if corrective action is warranted to prevent the recurrence of similar nonconforming products.
- Individual responsible for analyzing nonconforming product patterns to identify improvement opportunities.

Documented procedures are used to address nonconforming products. Information regarding nonconforming products should be tracked and documented as appropriate. The completion of a Nonconformance Report (NCR) is described here. These are general guidelines. This process may vary by industry.

1. The individual finding a nonconforming product should report it to the individual responsible for identifying nonconforming products. This person is often known as the Nonconformance Report Originator (NCR Originator).

2. Using a label, nonconformance tag, or appropriate means, the NCR Originator shall identify nonconforming products as soon as possible.

3. The NCR Originator shall ensure that nonconforming products are not inadvertently used or delivered to the customer.

4. If practical, nonconforming products shall be segregated to prevent their use. If segregation is impractical, other means must be taken to differentiate conforming and nonconforming products.

5. Care should be taken not to cause additional damage to nonconforming products. Appropriate storage and environmental controls shall be used.

(continued on next page)

6. The NCR Originator should complete fields 1–8 of the Nonconformance Report. These fields identify the nonconforming product. The NCR Originator should forward the NCR to the Responsible Manager, who will send it to the Corrective Action Request Coordinator (CARC) to enter into the NCR database.

7. The Responsible Manager or review team shall promptly evaluate the nonconforming product for impact on quality. The evaluation considers the following:
   - ❑ Analysis of design.
   - ❑ Impact on functionality and safety.
   - ❑ Alternative applications.
   - ❑ Costs and lead times for rework, repair, or return to supplier.

8. Based on the evaluation, the Responsible Manager or review team shall complete fields 9–17 of the NCR and send the form to the CARC for database updating. In doing this, they will have authorized one of the following dispositions:
   - ❑ Rework to original specifications.
   - ❑ Scrap.
   - ❑ Return to supplier.
   - ❑ Repair.
   - ❑ Regrade.
   - ❑ Use as is.

9. If the nonconforming product is submitted for rework or repair, the NCR should include:
   - ❑ A definition of required rework or repair.
   - ❑ Re-inspection criteria.

10. The responsible individual implements the disposition. For "use as is" or "repair" dispositions, a Deviation or Waiver is required. A Deviation or Waiver shall be approved prior to the next inspection activity or function. If the disposition is "use as is," the product is then used or integrated into the next phase of the assembly.

11. The Responsible Manager shall notify affected groups of the nonconforming product and its disposition. These groups may include:
   - ■ The supplier and/or customer.
   - ■ The purchasing, design, and operations groups.
   - ■ Review teams.

12. When required by customer agreement, the customer-signed Waiver or Deviation shall be controlled as a Quality Record.

13. If the product is reworked or repaired, it shall be re-inspected or re-tested. When the product is acceptable, the Responsible Manager signs field 18 of the NCR and forwards a copy of the NCR to the CARC.

14. The CARC ensures that all information is entered into the NCR database.

15. When warranted, the Responsible Manager shall ensure that corrective action is initiated to prevent the production of similar nonconforming products.

16. The Responsible Manager shall ensure that the nonconformance data is analyzed.

17. When warranted, the Responsible Manager shall ensure that preventive action is initiated.

18. The CARC shall analyze data relating to nonconforming products to determine root cause. When warranted, the CARC shall initiate preventive action.

(continued on next page)

 COPYRIGHT © GLENCOE/MCGRAW-HILL

Nonconformance Report

1. NCR No.

Identify Nonconformance

2. Originator: Phone: Date: Org:

3. Item Name: Qty:

4. Dwg/Part No: 5. Found During What Activity:

   Lot/Serial No:

6. Item Location:

7. Actual Condition:

8. Req'd Condition:

Disposition of Nonconformance

9. Authorized
   Disposition:
   - ☐ Rework To Original Spec.
   - ☐ Scrap
   - ☐ Return To Supplier For Replacement
   - ☐ Repair
   - ☐ Regrade
   - ☐ Use As Is

10. Disposition Rationale:

11. Cause Classification:
    - ☐ Design   ☐ Manufacturing
    - ☐ Supplier   ☐ Training
    - ☐ Other (*explain*)_____

12. Responsible Manager: Signature: Date:

13. ☐ Request Waiver   Waiver No:

14. Disposition Details:

15. Corrective Action Req'd   ☐ Yes   16. CAR No:
                               ☐ No

    Rationale (*if yes*):

17. Safety Function: Signature: Date:
    (*If Safety Is Affected*)

Disposition of Nonconformance

18. NCR Closed By: Signature: Date:

(continued on next page)

COPYRIGHT © GLENCOE/MCGRAW-HILL

MEETS THE FOLLOWING MSSC PRODUCTION STANDARDS

P1-K6	Document product and process compliance with customer requirements.
P1-K6A	Documentation of compliance is legible.
P1-K6B	Documentation of compliance is written in the appropriate format and correctly stored.
P1-K6C	Documentation of compliance is forwarded to the proper parties.
P1-K6D	Documentation is complete and "sign off" is obtained.
P1-K6E	Products are labeled appropriately for compliance or non-compliance.
P1-O4K	Knowledge of how to carry out non-compliance procedures.
P1-O6F	Skill in completing a compliance tag to indicate that the sub-assembly meets the customer requirements.
P4-K1D	Conformances to quality standards are properly assessed and documented.
P4-K4C	Materials are inspected against correct specifications.
P4-K4D	Materials that do not meet specification are correctly identified.
P4-K4E	Corrective action is taken on out-of specification material.
P4-K4F	Inspection results are properly documented.
P4-K4G	Inspection results are reported to correct parties.
P4-K6A	Appropriate corrective actions are identified and approvals received when needed.
P4-O1C	Skill in inspecting materials, and labeling and returning non-conforming materials.
P4-A14A	Write test records for quality control and non-conformance reports.
P4-A14B	Fill out reject material reports clearly and precisely.
P5-K3E	Communication is clear and relevant to quality.
P5-K3H	Communications are tracked and documented, as appropriate.
P5-K4A	Communication reflects knowledge of production requirements, levels, and product specifications.
P5-K4E	Communication is clear and relevant to production and products.
P5-O1C	Skill in completing a non-conforming product form to get approval for proper material disposition.
P8-K3	Determine appropriate action for sub-standard product.
P8-K3A	Quality procedures regarding sub-standard products are executed promptly within the defined quality systems.
P8-K3B	Decisions regarding sub-standard products are documented for future retrieval.
P8-K3C	Sub-standard product is appropriately processed.
P8-O4K	Skill in correctly tagging and segregating non-conforming material.
P8-O4L	Skill in investigating non-conformances (e.g., rejection tags) to determine root cause and recommend corrective action.

COPYRIGHT © GLENCOE/MCGRAW-HILL